Anatomy of the Honey Bee

ANATOMY OF THE
HONEY BEE

By R. E. SNODGRASS

Comstock Publishing Associates

A DIVISION OF CORNELL UNIVERSITY PRESS

ITHACA AND LONDON

Dedicated to the memory of

DR. EVERETT FRANKLIN PHILLIPS

Published in the United Kingdom by
Cornell University Press Ltd., London

First published in 1956 by Cornell University Press
First printing, Cornell Paperbacks, 1984

ISBN 978-0-8014-9302-7 (pbk. : alk. paper)
Printed in the United States of America

Cornell University Press strives to use environmentally responsible suppliers and materials to the fullest extent possible in the publishing of its books. Such materials include vegetable-based, low-VOC inks and acid-free papers that are recycled, totally chlorine-free, or partly composed of nonwood fibers. For further information, visit our website at www.cornellpress.cornell.edu.

FOREWORD TO THE
1984 PRINTING

ROBERT E. SNODGRASS (1875–1962) is best known to us in apiculture for this text and its predecessors *Anatomy and Physiology of the Honey Bee* (1925) and *The Anatomy of the Honey Bee* (1910). All told he wrote seventy-nine papers and books. His *Principles of Insect Morphology* (1935) is thought by many to be his masterpiece; it had much to do with his being awarded an honorary doctorate and is still the chief work in the field.

During most of his life Snodgrass was employed by the U.S. Department of Agriculture. He also worked, however, as a cartoonist, a sculptor, an artist, and a writer. Despite these many talents, his greatest interest was insects, their anatomy and how they functioned. The preparation of the major works on honey bees reflects his great concern for accuracy and precision. The 1925 book and this text are not revisions in the normal sense, for the early *Anatomy* was completely rewritten each time in an effort to provide a better insight into the functions of the honey bee. A strong believer in evolution, Snodgrass hoped that his works would aid others in gaining a better understanding of how things came to be what they are.

Snodgrass was also a popular lecturer whose presentations were vivid and practical. It is said that he would sometimes make blackboard illustrations with a piece of chalk in both hands.

None of his major works reveals Snodgrass the humorist, whose lively approach is evident in his "How the Bee Stings" (*Bee World* 14:3–6. 1933). That paper ends, discussing the sting: "The dagger is buried to the hilt in the flesh of the victim. The performance seen

v

enlarged in a cinema projection is highly instructive, though it becomes too vivid for real pleasure."

Late in his life he wrote, "My only regrets are the things I didn't do." He worked until the day he died believing that he had spent little time on hobbies and recreation. Acknowledging such single-mindedness, he claimed a similar interest in the hereafter, where he hoped to study the musculature of the angels in order to understand their flight mechanism.

ROGER A. MORSE

Ithaca, New York

PREFACE

THE butterfly is extolled for its beauty, the cicada for its song, the ant for its industry, but, of all the insects, the honey bee holds first place in our affections for the gustatory pleasure it affords us. Incidentally, also, the honey bee supports a large commercial industry; and probably for this reason the life of the bee is more intimately known than that of any other insect, and the anatomy of the bee has long been a subject of correlated interest.

First, it must be explained why the name of the bee appears in the title as two words, though "honeybee" is the customary form in the literature of apiculture. Regardless of dictionaries, we have in entomology a rule for insect common names that can be followed. It says: If the insect is what its name implies, write the two words separately; otherwise run them together. Thus we have such names as *house fly, blow fly,* and *robber fly* contrasted with *dragonfly, caddicefly,* and *butterfly,* because the latter are not flies, just as an *aphislion* is not a lion and a *silverfish* is not a fish. The honey bee is an insect and is pre-eminently a bee; "honeybee" is equivalent to "Johnsmith."

The present text is not a revision of the *Anatomy and Physiology of the Honeybee,* published in 1925; it is an almost entirely rewritten account of the anatomy of the honey bee, and the "honey bee" is understood to be not only the final winged adult of *Apis mellifera* L., but the whole series of forms the insect goes through from the germ cells and the egg to the adult, including the larva and the pupa. The field of physiology is now best left to the physiologists, and, in a text on anatomy, we may pass over the biology and behavior of the honey bee, since we now have on these subjects the recent books by von Frisch (1948), Ribbands (1953), and Butler (1954b). Because of the great increase during the last 30 years in our information about the structure and mechanisms of the bee, there is ample material for a book on these subjects alone.

The study of anatomy can be pursued merely from interest in the facts of animal structure, but it becomes more meaningful if we attempt to understand the mechanical and physiological reasons for anatomical facts. We may even admit that the object of anatomical study, except for purely taxonomic purposes, is to understand the animal as a working mechanism. There is also the intriguing problem for the anatomist of how the structure of any modern animal has come to be what it is, and how it has become so precisely adapted to the specific habits of each species; but here we enter into the realm of speculation and theory.

The fundamentals of insect anatomy were established by entomologists of the last century. The expansion of our knowledge of insect structure during the present century has made it inevitable that some of the ideas of homology which served well enough for the earlier entomologists should have to be readjusted to accord with what we now know from wider comparative studies. Unfortunately, it then follows that certain anatomical terms of long standing have to be discarded in favor of others that better express the facts. So, in the following chapters a number of terms current in apicultural literature will be replaced by others, and it is hoped that friends of the honey bee will not be offended by comparisons of the bee with such an inferior insect as the cockroach.

This book is appropriately dedicated to Dr. Everett Franklin Phillips, because it was due entirely to the interest of Dr. Phillips, then head of Apiculture in the old U.S. Bureau of Entomology, that the writer was enabled to do his first work on the anatomy of the honey bee, the results of which were published in 1910 as Bulletin 18 in the Technical Series of the U.S. Department of Agriculture. Most of the drawings then made are here reproduced for a third time. The McGraw-Hill Book Company has kindly turned over to the author the copyright on the *Anatomy and Physiology of the Honeybee*.

Information not otherwise available has been taken, by permission of the Princeton University Press, from Nelson's *Embryology of the Honey Bee* (1915). Also the writer is indebted to the Smithsonian Institution for material from his *The Skeleto-muscular Mechanisms of the Honey Bee* (1942). Then finally, for much new information contained in the following chapters credit must be given to numerous other writers, without whose work our knowledge of bee anatomy and histology would not be what it is today.

R. E. SNODGRASS

U.S. National Museum
Washington, D.C.

CONTENTS

Anatomy of the Honey Bee

FROM GERM CELLS
TO ADULT

IN THE narrowed upper ends of the ovariole tubules in the ovary of the queen bee are small masses of cells, very ordinary in appearance, yet from them will be produced all the new bees that will issue from the thousands of eggs laid by the queen throughout her long span of life. What gives these particular cells of the body their extraordinary powers cannot be explained. In many insects the *germ cells,* as the primary reproductive cells are called, are cells resulting from the cleavage of the egg nucleus that go into the posterior end of the egg, where by some special quality of the egg cytoplasm at this point they alone become endowed with the reproductive potentiality. After their ordainment the germ cells pass into the body of the embryo and become embedded in the mesodermal tissue of the forming ovary or testis. The origin of the germ cells in the bee has not been determined, since the germ cells of the bee are not distinguishable from mesoderm cells. However, Nelson (1915) aptly says:

It does not follow that they are of mesodermal origin, even though they seem to constitute a portion of the mesoderm, since it is not at all unlikely that the germ cells may be set aside at an early period in development, and afterwards migrate into the visceral wall of the mesodermal tubes, and that such a migration may take place unobserved.

In whatever manner the germ cells of an insect may have been differentiated from the prospective somatic cells, they are not replenished from somatic cells; they multiply in the gonad by division

1

and continue to do so as long as the insect retains its reproductive function.

The germ cells in the ovary are termed the *primary oogonia* and those of a male the *primary spermatogonia,* but we are concerned here only with the former, because every individual, male or female, begins as an egg. The cells formed from each oogonium pass down into the ovarial tubule, and here they undergo another differentiation by which one of them is destined to become an egg and the others to serve as food for the egg cell. The prospective egg cell is known as the *oocyte,* the food cells as *nurse cells,* or *trophocytes.* In the bee each oocyte is accompanied by about 48 nurse cells, all presumably derived from a single original oogonium. The oocyte increases greatly in size as it grows at the expense of its nurse cells, the entire substance of which is absorbed into the oocyte and becomes a nutritive material, called *deutoplasm,* or *yolk,* retained for the nourishment of the future embryo. The oocyte preserves its individuality as a cell, though it becomes enormously larger than any cell of the body. When it is fully formed in the lower end of the ovarian tube, the wall of the latter secretes a tough covering, or *chorion,* over the oocyte, which gives it the definitive shape of the mature egg. Though the oocyte has descended from an oogonium by a series of cell divisions, it will now, because of the large amount of yolk in its cytoplasm, no longer divide as a whole; subsequent divisions affect only the nucleus.

The mature oocyte enclosed in the chorion leaves the ovarial tubule and enters the oviduct, whence it goes into the vagina on its way to the outside. At about this time the nucleus undergoes its first division, but the result of this division is quite different from that of any previous division or of any that will follow, since the number of chromosomes in the nucleus is reduced to one-half of the usual number. By two successive divisions four nuclei are produced, one larger one will be the nucleus of the egg; the three smaller ones degenerate and are absorbed. This division process is called the *maturation* of the egg, because the oocyte thus becomes a mature egg, or *ovum,* and its nucleus is now ready to receive a spermatozoon from a drone. The egg of the honey bee, however, is capable of development without fertilization; it is potentially *parthenogenetic.* If it is not fertilized, it will usually develop into a male bee, or drone; a fertilized egg gives rise to a female, which will be a worker

2

or a queen according to the food given to the larva. Since the germ cells of the drone contain only half the number of chromosomes in a female cell, the union of a sperm nucleus with the maturated egg nucleus restores the normal female complement of chromosomes, and the fertilized egg becomes a female, while an unfertilized egg develops into a male. If the egg is inseminated, a few spermatozoa are admitted through a minute hole, the *micropyle*, in the anterior end of the chorion, and one spermatozoon unites with the egg nucleus, thus determining that the egg will develop into a female bee.

Whether an egg is to be fertilized or not is somehow determined by the queen herself at the time the egg is being laid. The queen has an apparatus (fig. 106 B) connected with the spermatheca (A, *Spt*), in which the sperm are stored, by which she can discharge a few spermatozoa on some eggs and withhold them from others. We cannot suppose that the queen uses any intelligence in the matter, and how she is able to regulate insemination of the eggs has been an outstanding mystery. The only guide she has would appear to be the relative size of the comb cells, since in general eggs deposited in small cells become workers and those deposited in larger cells become drones, but the queen cells are larger than the drone cells. However, even if the queen "perceives" the difference of cell size, this does not explain how she operates the sperm-ejection apparatus. The spermatozoa are carried to the egg in a secretion from glands connected with the spermatheca, and it is supposed that the spermatozoa are activated by the spermatic fluid. If discharge of the fluid is inhibited, therefore, the egg will be unfertilized. An idea advanced by Flanders (1950) is that the activation of the spermathecal glands is a reflex stimulated from the antennae, with which the queen examines a cell before oviposition and thus reacts to the relative size of the cell. Queen cells are constructed singly, the others in extensive groups, and Flanders suggests that the laying of a fertile egg in a queen cell results from a lag in the inhibition reflex when the queen goes from worker cells to a queen cell. For the same reason, he notes, the first eggs placed in drone cells are often fertilized eggs. The delivery of only a few spermatozoa to each fertilized egg involves a complex action of the spermathecal apparatus, which will be discussed in the chapter on the reproductive organs.

3

The egg of the honey bee is elongate oval in shape (fig. 1 A), about one-sixteenth of an inch in length and of a pearly white color. The chorion bears the imprint of the ovarial cells that secreted it. The egg is somewhat thicker at one end, which is to be the head end of the embryo, and is slightly curved with the convexity ventral and the concavity dorsal. The queen ordinarily places only one egg in each brood cell of the comb, attaching it by its smaller end to the inner wall of the cell, but sometimes she seems to be absent-minded and sticks an egg anywhere inside the cell, or even puts two or more eggs in the same cell. Since the cell, however, is designed to accommodate only one larva, the workers usually remove the extra eggs before they hatch.

The substance of the freshly laid egg consists of the cytoplasm of the original egg cell and of a much larger quantity of yolk material (fig. 1 A, *Y*), all surrounded by a delicate *vitelline membrane* (*Vit*), which is the true cell wall of the egg inside the chorion (*Cho*). Most of the cytoplasm is reduced to a meshwork enclosing the yolk globules, but a small cytoplasmic island in the anterior part of the egg contains the nucleus (*Nu*), and around the periphery is a denser cortical layer of *periplasm* (*Ctx*).

Development starts at once after the egg is laid. It begins with a division of the nucleus, followed by repeated divisions of the resulting nuclei until a large number are formed, which, embedded in small masses of the egg cytoplasm, scatter through the yolk as *cleavage cells* (fig. 1 B, *CCls*). The cleavage cells migrate outward and press into the cortical layer of cytoplasm; most of them merge into the latter forming beneath the vitelline membrane a continuous cell layer, which is the *blastoderm* (C, F, *Bld*). In the more primitive type of development among lower invertebrate animals the whole egg divides and the resulting cells divide until a globular mass of cells is formed, in which appears a central cavity, or *blastocoele* (fig. 65 A, *Bcl*). The same end result is accomplished by the insect egg, but by a different method because of the great amount of yolk in the egg, which fills the blastocoele at this early stage of development.

The blastoderm is at first of uniform thickness around the periphery of the egg, but soon the lower cells increase in size forming a distinct thickened tract along the convex ventral side of the egg, while the dorsal part of the blastoderm is reduced to a very thin

4

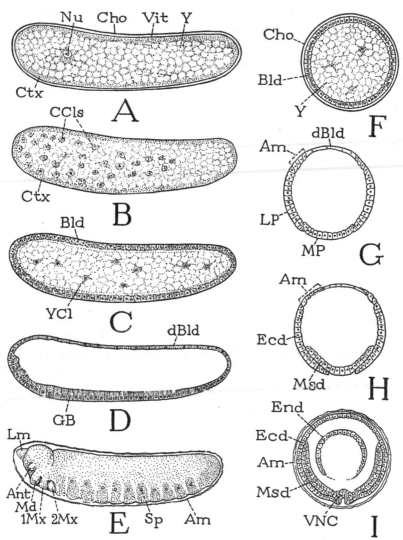

Fig. 1. Early development of the honey-bee egg and the young embryo (diagrams based on Nelson, 1915).

A, lengthwise section of egg before development, surrounded by chorion. B, cleavage cells formed by division of egg nucleus, scattering through the yolk and pressing into the cortex. C, cleavage cells mostly in the cortex, forming the blastoderm. D, blastoderm thickened on ventral side to form the germ band. E, young embryo 52 to 54 hours from beginning of development. F, cross section of egg in blastula stage. G, cross section of germ band differentiated into median plate and lateral plates. H, same, median plate becomes mesoderm, lateral plates ectoderm. I, cross section of young embryo in amnion.

For explanation of abbreviations see page 15.

5

layer of cells (fig. 1 D, *dBld*). The thick ventral tract is the *germ band* (*GB*), which is the beginning of the embryo. As development proceeds, the germ band spreads around the ends of the egg and extends upward on the sides at the expense of the dorsal blastoderm, which is soon reduced to a narrow strip along the back (G, *dBld*). Eventually the latter sinks into the yolk and the germ band closes over the dorsum, so that the young embryo becomes a cellular sac containing the yolk.

The germ band becomes divided by lengthwise grooves into a ventral *median plate* (fig. 1 G, *MP*) and a pair of *lateral plates* (*LP*). The median plate sinks into the yolk to become the *mesoderm* (H, I, *Msd*); the lateral plates then come together ventrally and constitute the definitive *ectoderm* (*Ecd*). Anterior and posterior ingrowths of the blastoderm form the *endoderm* (I, *End*) of the future alimentary canal, and mid-ventral thickenings of the ectoderm become the *ventral nerve cord* (*VNC*). Transverse external grooves mark the beginning of body segmentation, and outgrowing buds of the ectoderm form the rudiments of the future antennae, mouth parts, and legs. Details of the development of all these structures will be more fully described in the chapters on the organ systems. At about 53 hours from the start of development the embryo (E) begins to resemble a young larva. The embryo, however, is enclosed in a thin cellular membrane, the *amnion* (E, I, *Am*), which was not present at the beginning of development. The amnion is formed from narrow bands of cells along the upper edges of the lateral plates of the germ band (G, *Am*). The amnion cells grow together over the dorsal blastoderm (H, *Am*) and form anterior and posterior folds, which unite ventrally and finally enclose the embryo in a delicate sac. Shortly before hatching, the amnion is broken up by movements of the young larva. The larva will hatch at the end of three days from the deposition of the egg.

The young worker larva is 1.6 millimeters in length; it lies in a semicircle against the bottom of its cell. It is now provided by the attendant nurse bees with an abundance of nutritious gland-produced food, known as *royal jelly*, or *bee milk*, which is rich in fats and albumens. The queen-to-be larva gets royal jelly throughout its lifetime, but after about the third day the worker and drone larvae are put on a diet of nectar (or honey) and undigested pollen. The larvae grow rapidly: according to Nelson,

Sturtevant, and Lineburg (1924) a worker larva within 4½ to 5 days increases its initial weight 1,500 times. At the end of the fourth day it has grown so large that it is now coiled in a circle at the end of its cell (fig. 2 A).

Attending to one larva requires a great deal of work on the part of the nurse bees. It is said by the writers quoted above that during the eight days from the time the egg is laid until the mature larva is sealed in its cell, the attendant bees average about 1,300 visits a day. The larval workers, queens, and drones moult at approximately 24-hour intervals during the first four days of life. The fifth (and last) moult of the worker larva takes place at the end of the eighth day, that of the queen at the end of the seventh day, but the drone larva does not undergo its last moult until the end of the eleventh day. Some time before the last moult the larval cell is sealed by the worker bees with a cap of wax, and the larva now gorges itself on the remains of its food. This done, the larva spins a silk cocoon about itself, evacuates the refuse in its alimentary canal, and stretches out on the floor of its cell with its head toward the outer end. In this position it undergoes its last larval moult. The discarded skin, Bertholf (1925) says, "is pushed back to the base of the cell where it mixes with the yellow feces, given off by the larva during or just after the spinning of the cocoon, and the whole sticks to the base of the cell as a yellow flake."

The bee larva has little resemblance to an adult bee; in fact, it can hardly be regarded as a young bee, since to become a bee it has to be almost wholly reconstructed in the pupal stage. The larva is a wormlike grub fitted to live in a cell of the hive comb, where it is fed by adult attendants. It has simple mouth parts suitable for its own feeding, but it has no external antennae, legs, wings, reproductive organs, or sting, and no eyes. In these characters the bee larva is not different from the larvae of other beelike or wasplike Hymenoptera, but it is quite different from the larvae of the related sawflies, which live in the open and are equipped with functional legs because they must take care of themselves and find their own food as soon as they are hatched. It is probable, therefore, that the legless hymenopterous larvae were evolved from some caterpillarlike larva, and it must be noted that none of the former, not even the bee larva, is *truly* legless, since there are

7

budlike rudiments of legs growing in pockets of the epidermis beneath the thoracic cuticle (fig. 2 C, *L*). Also, such larvae are not wingless since they have internal wing buds (*W*) similar to the leg rudiments, and finally they have antennae growing beneath the cuticle of the head and, in later stages, rudiments of an ovipositor near the end of the abdomen (*D*). Fundamentally, therefore, the larva is equivalent to the nymphal stage of an insect without metamorphosis.

The bee larva is externally wingless because the bees belong to the great group of endopterous insects, in all of which the wings grow internally during the larval stages. The suppression of external legs, however, is limited among the Hymenoptera to the members of the clistogastrous group, in which the larvae are parasitic, or live in plant tissue or in the protection of nests. The "legless" condition of the bee larva, therefore, was not evolved specifically as an adaptation to life in a comb cell; rather, because the honey bees inherited legless larvae from their progenitors, they were able to rear their young in cells of the comb. If their larvae had functional legs and could freely run about, they would be a great nuisance in the hive, and there would be no keeping them at home. The hive bees, therefore, owe much of their way of living to the fact that their larvae can be stored in cells and kept there until they come out as fully fledged adult bees with instincts for doing only what they should do. The bee society has no problems of juvenile delinquency.

At each of its first four moults the honey-bee larva comes out of its old skin with little external change other than an increase in size, but when the skin is shed a fifth time there is revealed a new creature having no likeness to the larva. This is the *pupa* (fig. 2 I). The change of form, however, took place within the larval skin some time before the skin was shed, and in this stage the insect is known as a *propupa* (E). The propupa itself undergoes developmental changes (G, H) before it becomes a fully formed pupa (I). Bertholf (1925) says the bee larva "passes gradually and without moulting" into the prepupal stage, but the fact that the propupa is freely enclosed in the larval skin (E, *LCt*) shows that moulting has already taken place. *Moulting*, properly defined, is the separation of the outer layer of the cuticle by a dissolving of the inner layer. It is only when the epidermis is thus free of the old cuticle that it can form a new cuticle and begin a new phase of growth, which, after the last larval moult, produces the propupa. The final emergence of the insect from the larval skin is its *ecdysis* (coming out). The so-called propupa, then, is merely the pupa in

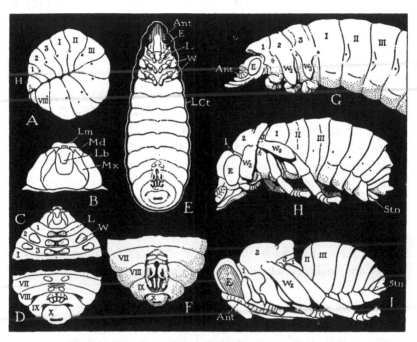

Fig. 2. External changes during metamorphosis from larva to pupa.
A, mature larva coiled in cell. B, ventral surface of larval head. C, ventral surface of head and thorax of mature larva, leg and wing buds everting beneath larval cuticle. D, terminal segments of worker larva, ventral, showing rudiments of sting. E, early stage of propupa in larval cuticle. F, terminal segments of propupa with rudiments of sting more advanced. G, lateral view of propupa. H, late stage of propupa just before shedding larval cuticle. I, pupa after ecdysis.
For explanation of abbreviations see page 15.

its early developmental stages within the moulted skin of the larva. Insects such as the house fly and its relatives pass the entire pupal period inside the moulted but unbroken third larval skin, called a *puparium*.

In the earlier stages of development the propupa (fig. 2 E, G) begins to take on adult characters in the head and thorax, but the abdomen is still larvalike and is not constricted from the thorax. Compound eyes are now present, the mouth parts have something

9

of the adult structure, the wings and legs are fully exposed, the legs though small are distinctly jointed, but the wings (G, W_2, W_3) are merely flaps hanging down from the back; the rudiments of the sting (F) have increased in size since the last larval stage (D). The propupa develops rapidly, and shortly before ecdysis (H) it has taken on much of the form of the adult bee, though still the mouth parts are not fully developed, the wings are short pads, and the definitive constriction between thorax and abdomen is not yet present.

When the larval skin is finally cast off, the insect appears in a form that is unquestionably that of an adult bee (fig. 2 I), and it is now known as a pupa. The wings are still small, but the head, antennae, mouth parts, thorax, legs, abdomen, and sting all have adult characters. A deep dorsal constriction separates the thorax from the abdomen, but it is to be seen that the constriction is in front of the second segment of the abdomen of the propupa (H, *II*), and that the first segment (*I*) has been joined to the thorax (I, *I*). This feature is characteristic of all adult clistogastrous Hymenoptera, in which the definitive thorax contains four segments, the included abdominal segment being known as the *propodeum*.

The pupa does not grow or change in shape during the rest of the pupal period. Its cuticle is hardened, and the epidermis must therefore wait until another moult removes the cuticle before it can form the adult integument with its furry covering and give the final touches to the appendages. Inside the pupa, however, great changes are in progress, involving a reconstruction of the musculature and the alimentary canal, changes in most of the other organs, and the development of the organs of reproduction. Most of these transformation processes begin in the propupa and some in a late stage of the larva, but they will be described in subsequent chapters. The pupa is a reconstruction stage of the insect, in which business does not go on as usual; larval organs are being broken down and adult organs formed. Externally there is no sign of all this, and the shape of the pupa is that of the adult in order that the newly formed muscles and other internal parts may be fitted to the form and dimensions of the imago. The pupal stage of a worker bee lasts for nine days, that of a queen five days, and that of a drone eight or nine days (Bertholf, 1925). Before ecdysis,

10

however, the cuticle has been loosened, i.e., moulted, and within it the adult form is finally completed, including the full development of the wings and the external clothing of hairs. During all its period of development the pupa has been sealed in the cell of the comb in which the larva was reared, so that the adult on shedding the pupal skin has still to liberate itself by gnawing through the cell cap.

Once emerged from the cell, the adult bee has only to stretch its legs, extend its antennae, allow its wings to unfold and stiffen and its fur to dry. Then, if it is a worker, it is ready to assume its appointed duties in the hive. Should it be a queen, however, and be so unfortunate as to be preceded by another queen already emerged, it is likely to be liquidated in its cell by the first queen. A drone has little to do in the hive except to solicit food and await the time when a virgin queen will require his services. Two or three days after mating the queen begins laying eggs.

Inasmuch as the adult bee is the principal subject of the following chapters, we need give little attention here to the details of its structure. In its natural condition the body of the bee is so densely furry that much of its structure is obscured. The hairs can be rubbed off with a brush, but they are more easily removed by placing the bee in melted paraffin wax, allowing the latter to harden, and then carefully breaking it away. Most of the hairs will come off embedded in the wax, leaving the body of the bee fairly bare and in good condition for study (fig. 3). We may now see clearly that the three anatomical sections of the bee are sharply separated. The thorax is the sustaining part of the insect and is the locomotor center since it bears the legs and wings. The head is supported on the thorax by a narrow neck, and the abdomen is attached by a slender petiole. As already noted in the mature pupa (fig. 2 I), the constriction separating the abdomen from the thorax is between the primary first and second abdominal segments of the larva (A) and the propupa (G, I, II). The transfer of the first abdominal segment to the thorax gives increased space in the latter, which allows a great lengthening of the dorsal muscles of the wing mechanism, and the shortened petiolate abdomen probably has more freedom of movement than it would have if joined directly to the third thoracic segment. At least we can be pretty sure that whatever special structures an insect has are for its own advantage.

11

The elongate proboscis allows the bee to draw nectar from the depths of flower corollas, the broadened hind legs of the worker serve for the transport of pollen, the extra segment in the thorax gives more space for wing mucles, the petioled abdomen acquires

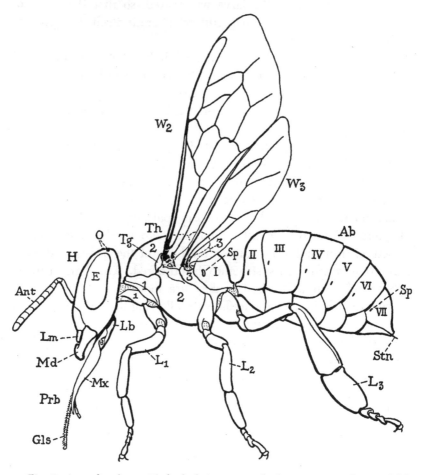

Fig. 3. A worker bee with body hairs removed, showing appendages of left side.

For explanation of abbreviations see page 15.

freedom of movement for the use of the sting and allows the queen more easily to insert her abdomen into the brood cells of the comb for egg laying.

The anatomy of an animal is merely the machinery by which the animal lives, and we can now explain most of the mechanisms by which the worker bee carries on her activities within and without the hive. Yet the mystery remains of how the bee knows how and when to use the various tools and gadgets with which she is provided. For the perception of her surroundings the bee is amply endowed with sense organs, but how her sensory perceptions elicit the appropriate reactions in the motor mechanism is a question beyond the field of the anatomist—at least until the anatomist explores the intricacies of the central nervous system, and even then there is no assurance that the "intelligence" of an insect will be understood. The "know-how" of an insect is still a baffling thing, and for the present we still have only the blanket term "instinct" to cover our ignorance.

The caste system is rigidly fixed in the bee society; there is no overlapping of functions between drones, queen, or workers, except that under certain conditions a worker can produce eggs. The drone is only a functional male, but the queen and workers are differentiated females. The evidence has seemed pretty conclusive that the differences between workers and queen are produced by differential feeding; the queen gets the more nutritious diet of royal jelly through her entire larval life, while larvae destined to be workers get little of it after their first three days.

Though young female larvae are actually reared by the attendant bees into either workers or queens, the problem of accounting for the difference on a purely nutritional basis is complicated by the fact that, while the queen acquires highly developed ovaries, she lacks all the special features of the worker, such as specialized mandibles, food glands, wax glands, scent gland, the pollen-carrying apparatus of the hind legs, and the complex worker instincts. By contrast with the honey-bee queen, the bumble-bee queen (*Bombus*) retains all the worker characters and instincts in combination with well-developed ovaries; she is both a worker and a reproductive female. If, therefore, the *Apis* queen has been derived from an ancestor like the *Bombus* queen, the original female characters of the species were somehow apportioned between the two female castes. The problem of caste determination then is not one of rich or poor feeding, but one of bilateral inhibition.

In the literature on the honey bee much has been written about royal jelly, its composition, and its influence in differentiating the queen from the worker (see Haydak, 1943, Willson, 1955).

Experiments on the artificial feeding of unlimited amounts of royal jelly to larvae throughout their lifetime, such as those of Weaver (1955), have not produced queens, and Weaver suggests that the workers must add some other substance to the queen larva's diet that brings about the suppression of the special worker characters.

It is the relative permanency of the honey-bee colony, in contrast to the temporary life of a bumble-bee colony, that has allowed the *Apis* queen to give up the labors of a worker and devote herself entirely to the matter of procreation. Consequently, the queen does not need the worker's tools; but, being a female, she cannot be relieved of them by heredity. Therefore, a young female larva is reared individually in a special cell to be a queen. The method by which the transformation is brought about, however, is still the secret of the worker bees, but their magic formula must contain a specifically inhibitory ingredient as well as an excess of some ovarial stimulant. It has been suggested that an inhibitory hormone may be produced by the developing ovaries of the queen, but no such hormone has been demonstrated to be present.

A colony of bees is a biological unit. It may be likened to a complex organism in that the individual bees and their caste groups are comparable to the cells and organs of an animal which all work together for the good of the whole. There is this difference, however, that in the animal organism all activities are regulated from governing centers in the nervous system, while in the bee colony each individual bee is self-regulating. The bee colony, in fact, is a perfect example of communism fully achieved. Individualism has been so completely suppressed that no bee acts or even thinks as an individual. So perfectly is each member conditioned to the work it has to do that no dictator is needed. The system probably has worked for several million years and could continue unchanged for millions more. Of course, it must be assumed that insects have always been guided by instinct and never did think for themselves or have individualistic aspirations. Insects, therefore, are ideal subjects for communal organization, and we, as less perfect social animals, can learn from the bees, the ants, and the termites more than we should ever care to put into practice.

Explanation of Abbreviations on Figures 1–3

Ab, abdomen.
Am, amnion.
Ant, antenna.

Bld, blastoderm.

CCls, cleavage cells.
Cho, chorion.
Ctx, cortical cytoplasm, periplasm.

dBld, dorsal blastoderm.

E, compound eye.
Ecd, ectoderm.
End, endoderm.

GB, germ band.
Gls, glossa.

H, head.

I, propodeum.
I-X, abdominal segments.

L, leg.
Lb, labium.
LCt, larval cuticle.
Lm, labrum.
LP, lateral plate of germ band.

Md, mandible.
MP, middle plate of germ band.
Msd, mesoderm.
Mx, maxilla.
1Mx, first maxilla.
2Mx, second maxilla.

Nu, nucleus.

O, ocelli.

Prb, proboscis.

Sp, spiracle.
Stn, sting.

Tg, tegula.
Th, thorax (*1, 2, 3*, thoracic segments).

Vit, vitelline membrane.
VNC, ventral nerve cord.

W, wing (*W$_2$*, *W$_3$*, mesothoracic and metathoracic wings).

Y, yolk.
YCl, yolk cell.

II

THE BODY WALL
AND THE MUSCLES

AN INSECT is a living machine; no other animal is provided with so many anatomical tools, gadgets, or mechanisms for doing such a variety of things as is a winged insect. The movable units are parts of the body wall, including the appendages; the motor elements are the muscles attached on the body wall. The insect mechanism, however, is very different from any man-made machine. There is no single source of power, such as the engine of a motor car; each of the movable parts of an insect has its own motor unit in the form of attached muscles. All action is then controlled and co-ordinated from the central nervous system, just as if the driver of a car had not only to do the steering but to operate separately the engine and the movement of each wheel. Since a muscle exerts force only by contraction, a movable part is usually provided with antagonistic pairs or groups of muscles, but in some cases a reverse movement results from the elasticity of the body wall. Inasmuch as the distinctive characters of an insect are shown in its external structure, topographic maps of the skeleton may be sufficient for taxonomic purposes, but if we wish to understand the insect as a working mechanism, there is little point in studying either the skeleton or the muscles alone.

THE BODY WALL

The body wall is a highly important part of the insect organization not only because it forms a protective integument, but because, by the hardening of its outer part, it becomes an exo-skeleton, the structure of which determines the motor activities of the insect. The essential part of the body wall is a cellular epithelium (fig. 4 A, *Epd*) representing the part of the embryonic ectoderm that remains at the surface. This cell layer of the arthropod integument is best termed the *epidermis,* since it corresponds with

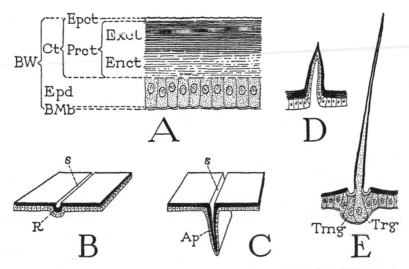

Fig. 4. The body wall.

A, diagram of body-wall structure, with nomenclature. B, a shallow inflection of the body wall forming an external sulcus (*s*) and an internal apodemal ridge (*R*). C, a deep inflection forming a platelike apodeme (*Ap*). D, a multicellular external process of the body wall. E, a unicellular external hairlike process, or seta.

For explanation of abbreviations see pages 29–30.

the epidermis of a vertebrate animal, though it has commonly been called the "hypodermis" by entomologists. A true hypodermis is not present in insects or other arthropods. The epidermis is lined internally by an apparently noncellular *basement membrane* (*BMb*) and is covered externally by a *cuticle* (*Ct*). The epidermis is responsible for the formation of the cuticle, but the cuticle becomes anatomically the important part of the body wall. It varies in thickness, structure, and composition on different parts of the body. In the bee larva the cuticle is soft and very thin as compared with the thickness of the epidermis; in the adult the cuticle becomes greatly thickened, and the epidermis is usually reduced to a thin epithelium.

The cuticle was formerly regarded as a chitinous secretion of the epidermal cells. It is a secretion and it does contain chitin, but it has other important constituents, chief of which is a protein. The cuticle includes two distinct layers, a thick inner layer containing chitin, termed by Richards (1951) the *procuticle* (fig. 4 A, *Prct*), and a very thin, nonchitinous surface layer, or *epicuticle* (*Epct*). Where the cuticle is hardened, or sclerotized, the procuticle is differentiated into an outer, darker zone of dense *exocuticle* (*Exct*) and a pale, laminated inner zone of softer *endocuticle* (*Enct*). The important constituent of the exocuticle is protein, and the sclerotization of the exocuticlar zone has been shown by Pryor (1940) to be the result of the "tanning" of the protein by a dihydroxyphenol, producing a substance called *sclerotin*. "A similar reaction," Pryor says, "takes place in the epicuticle, which is secreted as a protein membrane," and later becomes impregnated with fatty (lipid) materials. It is the sclerotized exocuticle that forms the plates, or *sclerites*, of the body wall; the flexible areas are merely nonsclerotized parts. It is now well known that the soft parts of the cuticle contain a larger percentage of chitin than do the sclerites, in which protein in the form of sclerotin predominates.

The cuticle forms not only the exoskeleton of the insects, but also various internal structures that result from ingrowths of the cuticula in formative sheaths of epidermis. Such structures, in general termed *apodemes*, take the form of ridges (fig. 4 B, *R*), plates (C), arms, or muscle tendons. On the other hand, outgrowths of the body wall form various external structures, some of which are multicellular spurs or spines (D) though the majority are unicellular hairlike setae (E). In a wide sense we should include as ingrowths of the body wall ectodermal glands and their ducts, tracheae, and the outlet ducts of the reproductive organs; and in a last analysis antennae, mouth parts, legs, wings, and external genital organs are outgrowths.

Sclerotized regions of the cuticle are often divided into secondary areas by grooves, or *sulci* (fig. 4 B, *s*). Such grooves are commonly called "sutures" in entomological terminology, but a suture properly so called should be a line where primarily separate parts have grown together, the term *suture* being derived from the Latin *suo*, "sew." The only true anatomical sutures are those made by surgeons. The external grooves of the insect skeleton are simply the lines along which the cuticle has been inflected to form internal ridges (*R*), which serve to strengthen the skeleton in the manner of the rafters of a house, or to give attachment to muscles. The apodemal ridges, therefore, are the important structural features, and not the "divisions," or resulting "plates," of the external surface, except when the latter are separated by flexible conjunctivae.

The entire back of an animal is the *dorsum,* the entire undersurface is the *venter.* The dorsal and ventral plates of a segment, according to the terminology to be followed here in describing the skeleton of the bee, are respectively the *tergum* (or *notum*) and the *sternum.* Many entomologists, however, call the dorsal segmental plates "tergites" and the ventral plates "sternites," but, since the suffix -*ite* in zoology means "a part of," this usage leaves no terms available for secondary subdivision of a back plate or a ventral plate. It may be noted that we never see such adjectives as "intertergital" or "tergitosternital," which consistency would require.

THE MUSCLES

The muscles of insects consist of fibers rather loosely associated in bundles that have no retaining sheath, or *perimysium.* They are therefore often branched at the ends, and what is a single muscle in one species may in another be split up into two or more muscles having a common point of origin. Insect muscles are of the striated type of structure, in which light and dark bands alternate (fig. 5 A, J, Q). Each light band is cut into halves by a *telophragma* (Z), and the parts of the fiber between the telophragmata are known as *sarcomeres.* The sarcomere thus consists of the dark band (*Q*) and of a light *end disc* (*E*) at each end; the dark band itself is traversed by a faint *mesophragma* (*M*), and the end disc may have a narrow dark *accessory disc* (*N*). Each fiber consists of a bundle of numerous parallel *fibrils,* or *sarcostyles* (*Fbl*), ensheathed in a delicate *sarcolemma* (*Sarl*). Since a muscle fiber is formed by the end-to-end union of primitive muscle cells, or *myocytes,* each fiber has many nuclei, which may be superficial beneath the sarcolemma or embedded in the body of the fiber.

The somatic muscles are attached on the cuticle of the body wall or on cuticular ingrowths such as apodemes and tendons. The

19

attachment is by means of bundles of fine threads known as *tonofibrillae* (fig. 5 B, E, *Tfbl*), which themselves appear to be of a cuticular nature (D). The methods by which the tonofibrillae are formed and become attached to the muscles, however, have

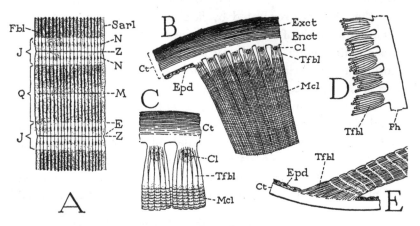

Fig. 5. Muscles, their structure and attachments.
A, diagram of structure of a muscle fiber, with nomenclature of its parts. B, fibers of a wing muscle attached by tonofibrillae to cuticle of tergum. C, wing-muscle fibers showing continuity of tonofibrillae with sarcostyles. D, tonofibrillae of wing muscle attached to a phragma. E, attachment of leg-muscle fibers.
For explanation of abbreviations see pages 29–30.

still not been clearly elucidated. Yet it seems that the tonofibrillae must be products of the epidermal cells where the forming muscle comes in contact with the epidermis. At some places apparent remnants of epidermal cells may be seen within groups of tonofibrillae (B, C, *Cl*). Oertel (1930) says that the attachment of the muscle to the cuticle in the pupa of the honey bee is preceded by an elongation of the epidermal cells at the point where the attachment is to be, and that the cell cytoplasm forms into filaments that fuse with the muscle. It seems reasonable to assume that the production of tonofibrillae is induced by contact of the newly forming muscle with the epidermis. Each tonofibrilla appears to be continuous with a single fibril of the muscle fiber (fig. 5 C).

The Larval Muscles— The muscles of the honey-bee larva have the structure characteristic of arthropod muscles in general and of most insect muscles. The striated fibrillar core of each fiber

20

(fig. 6 B) is surrounded by an outer layer of sarcoplasm (*Sar*), in which are contained the nuclei (*Nu*) beneath a limiting sarcolemma (*Sarl*). Where the fiber is attached to the body wall (A), the sarcoplasm of the fiber merges into the cytoplasm of the epi-

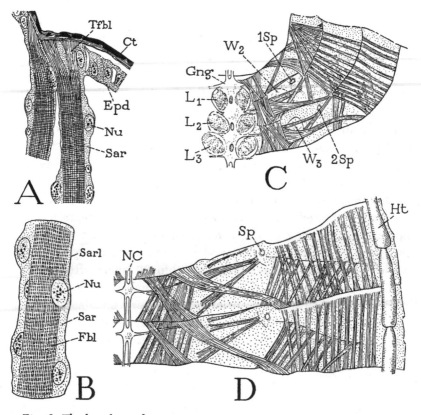

Fig. 6. The larval muscles.

A, section of a larval muscle attached to cuticle. B, surface view of a larval-muscle fiber. C, musculature of right half of larval thorax (from Nelson, 1924). D, muscles of right half of two abdominal segments of larva (from Nelson, 1924).

For explanation of abbreviations see pages 29–30.

dermis and the sarcolemma becomes continuous with the basement membrane.

The somatic muscles of the bee larva, as shown by Nelson (1924), are strictly metameric in their arrangement and conform closely to a definite pattern in all the body segments (fig. 6 C, D), except that in the thorax (C) they are somewhat less regular than in the abdomen (D), and in the tenth abdominal segment the musculature is reduced to a few retractors of the anus. The abdominal musculature (D) includes longitudinal and oblique dorsal muscles, tergosternal and oblique lateral muscles, and longitudinal and oblique ventral muscles. In the head are muscles of the mouth parts and the labrum, but the antennal and leg rudiments of the larva have no muscles. The larval musculature as developed in the embryo is clearly designed to serve the wormlike larva. Although in a general way the abdominal musculature of the larva (fig. 6 D) is similar to that of the adult (fig. 54 A), the larval thoracic musculature has little to suggest that of the winged adult. Moreover, the structure of the adult muscles is quite different from that of the larval muscles. The somatic muscular system, therefore, is almost entirely reconstructed in the pupa for the purpose of the adult bee.

Metamorphosis of the Muscles— The muscles of the bee larva are said by Oertel (1930) to show no sign of disintegration five hours before sealing of the comb cell, but there are present at this time, in addition to the larval muscle nuclei, groups of small nuclei lying close to the muscles. Within 30 hours after sealing, these external nuclei have increased in numbers, and it is now evident that they are the nuclei of the future imaginal muscles. The pupal metamorphosis of the muscles includes the formation of new imaginal muscles having no representatives in the larva, the reconstruction of larval muscles into imaginal muscles, and the complete destruction of larval muscles to be replaced by imaginal muscles.

Imaginal muscles newly formed in the pupa include the muscles of the antennae, the legs, and the sting, since the rudiments of these appendages in the larva have no muscles. The myoblasts of the leg muscles, according to Oertel, appear in a larva five hours before sealing as groups of small spindle-shaped cells near the leg bases. The myoblasts presumably are adventitious mesoderm cells. In the pupa they elongate, and rows of them unite, forming fibers with axial nuclei. At 70 hours after sealing, the fibers begin attaching to the epidermis.

The reconstruction of larval muscles into imaginal muscles takes place principally in the abdomen. In a larva of 10 hours after sealing, Oertel says, small myocytes appear among the longitudinal abdominal muscles, they increase in numbers and penetrate into the tissue of the disintegrating larval muscles. At 60 hours after sealing, a large part of the muscles of the larva has disappeared, leaving only a mass of myocytes. The myocytes, growing at the expense of the muscle substance of the larva, become arranged in rows to form new fibers with axial nuclei.

A total destruction of the larval muscles takes place in the thorax. At a late larval stage, according to Oertel, the larval thoracic muscles have been completely histolyzed, and the thorax is filled with a mass of myocytes and nuclei. The free myocytes that appear in the body are supposed to be undifferentiated mesoderm cells, but apparently their exact origin has not been observed. At 10 hours after sealing, the myocytes are arranged in rows; 25 hours later muscle formation is in progress; and at 70 hours the muscles are attached on the thoracic wall. The largest muscles of the thorax are the indirect muscles of the wings, and these muscles, as will presently be shown, have quite a different structure from the other muscles of the adult bee.

The Adult Muscles— The muscles of the adult bee are all either reconstructed from larval muscles or they are newly formed in the pupa. Not only do they differ in structure from the larval muscles, but they present two different types of structure of their own, one being that of the body muscles in general, the other characteristic of the indirect wing muscles.

The structure of an ordinary body muscle is well shown in a leg muscle. A fiber of one of these muscles in surface view (fig. 7 B) looks like a larval muscle except that there is but a scant surface layer of sarcoplasm and no nuclei are to be seen. A lengthwise section of a fiber (C) shows that the nuclei lie in an axial core of sarcoplasm, and a transverse section (D) reveals that the fibrils (*Fbl*) are arranged in the form of lamellae radiating outward from the fiber axis, many of them splitting into two or more leaves at the outer ends. Fibers of this type are characteristic of the body muscles not only of the adult bee but of the adults of all the higher Hymenoptera, as well as of the flies of the order Diptera and of some other insects.

The great indirect muscles of the wings that occupy so much space in the thorax (fig. 48) are distinguished in appearance by their brownish color in contrast to the whiteness of the other

23

muscles and in structure by the thickness and individuality of the sarcostyles of the fibers. These muscles in the bee have been elaborately described and discussed by Morison (1928a) and need but brief treatment here. An exhaustive study of the anatomy and

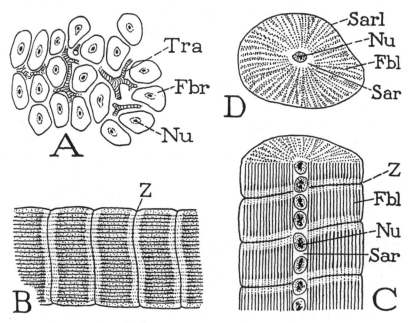

Fig. 7. Structure of adult muscles other than indirect wing muscles.
A, cross section of leg muscle, fibers interwoven by tracheae but otherwise unconnected. B, surface view of leg-muscle fiber. C, lengthwise section of leg-muscle fiber. D, cross section of same.
For explanation of abbreviations see pages 29–30.

histology of insect flight muscles has recently been made by Tiegs (1955).

The fibers of the indirect wing muscles of a worker bee are from 90 to 160 microns in diameter. They are closely packed (fig. 8 B), but in a cross section (A) their polygonal forms are seen to be neatly fitted together with spaces between them which admit numerous tracheae (*Tra*) that penetrate among them. The surface of each muscle is sheathed in a network of tracheal branches and small irregular air sacs (C), from which numerous parallel tracheae arise and go directly into the body of the muscle and ramify over the surfaces of the fibers. The tracheae finally branch into tracheoles, and, according to Morison, "it can be seen that every sarcomere on the exterior of a fresh fiber has at least one tracheole passing over it, and many of the tracheoles can be seen to pass into the fiber between the fibrils." The indirect wing muscles

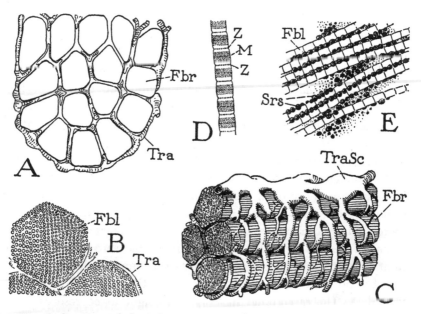

Fig. 8. Structure of the indirect wing muscles.
A, cross section of lower part of a dorsal muscle, showing large fibers interlaced by tracheae. B, cross section of fibers, showing fibrils (*Fbl*), or sarcostyles. C, group of fibers with surface tracheae and air sacs. D, part of a sarcostyle. E, group of sarcostyles separated, showing darkly stained sarcosomes (*Srs*).
For explanation of abbreviations see pages 29–30.

of the bee belong to the mesothorax, and most of their tracheae come from large tracheal trunks arising at the first spiracles, which are mesothoracic in origin, though in the adult they appear to belong to the prothorax.

When a fiber of one of the indirect wing muscles is dissected in water or glycerine it breaks up readily into a multitude of threads, which are very coarse compared with the delicate fibrils often barely visible in the other muscles. Yet, since these elements of the wing muscles are not further divisible, they are evidently the true fibrillae, or sarcostyles, of the fibers. The structure of a sarcostyle is in many ways strikingly similar to that of a fiber itself. Each fibril is a cylindrical thread (fig. 8 D) about 3 microns in diameter, extending through the entire length of the fiber and surrounded by a delicate retaining membrane. The sarcostyle is crossed by alternating light and dark bands, and the light bands are bisected by distinct telophragmata (Z), cutting the sarcostyle into sections like the sarcomeres of a fiber. Through the middle of each section is a faint dark line, presumably the edge of a mesophragma (M). The light and dark discs are so evenly aligned from one sarcostyle to another that they give the appearance of transverse striation on the fiber as a whole. The telophragmata of neighboring sarcostyles coincide so closely in position as to suggest that there is some invisible connection between them bridging the interfibrillar spaces.

A second distinctive feature of the indirect wing muscles is the great number of *sarcosomes* between the sarcostyles (fig. 8 E, Srs). The sarcosomes are oval, protoplasmic bodies mostly lying in continuous rows between the sarcostyles, regularly two to each sarcomere, but in dissections they break up into masses of irregular grains of various sizes. In fresh specimens the sarcosomes have a faint brownish tint that perhaps gives the brown color to the indirect wing muscles; in stained specimens the sarcosomes become very dark. The sarcosomes have been thought to be reserve food material for the muscles, their abundance in the wing muscles being attributed to the need of these muscles for a constant source of energy to sustain their rapidity of contraction and their long-continued activity when the insect is in flight. However, from a study of the structure, histochemical reactions, spectroscopic properties, chemical composition, and enzymatic contents of the sarcosomes of flies, Watanabe and Williams (1951) conclude that the sarcosomes of the wing muscles are equivalent to the mitochondria of other cellular tissues, which in general are minute granules, or short rods or filaments in the cell cytoplasm. These same writers (1953) find that in newly emerged flies, *Drosophila* and *Phormia*, the sarcosomes are about one micron in diameter, but during the first week of adult life they grow rapidly in diameter and become constant in size at 2.5 microns. The sarcosomes evidently, therefore, are not "consumed" by the sarcostyles. According to the recent work by Edwards and Ruska (1955) on the muscles of beetles, the indirect muscles of flight are penetrated by tracheoles from the surface tracheae, which run between the sarcostyles and end in intimate associations with the sarcosomes. The sarcosomes, or huge mitochondria of the indirect flight muscles, are therefore explained by these writers as the loci of oxidative enzymes for the sarcostyles, their great size being correlated with the activity of the flight muscles and the consequent need for high consumption of oxygen. The chemistry of the muscles of the honey bee has been discussed by Morison (1928b).

Nuclei are difficult to see in the indirect wing muscles because they so closely resemble the sarcosomes. Morison (1928a), however, says that by appropriate staining methods the nuclei can be clearly distinguished among the sarcosomes between the sarcostyles.

In the adult honey bee, including the three castes, there are over 200 pairs of somatic muscles. The individual muscles and their functions have been described by Morison (1927) and by the writer (1942). A complete list of muscles, therefore, will not be included in the present text, but due attention will be given to those muscles that operate the various mechanisms of the bee.

BODY SEGMENTATION

Segmentation of the body is determined first by the attachment of the longitudinal muscles on the body wall, but it is secondarily modified to preserve freedom of movement when the integument becomes sclerotized. In the soft-skinned larva of the bee the intersegmental lines are merely circular inflections of the body wall on which the longitudinal muscles are attached (fig. 9 A, isg). The coincidence of muscle attachments and intersegmental grooves suggests that body segmentation has resulted from the attachment of the muscles at successive intervals on the body wall. Segmentation of this simple type, therefore, may be termed *primary segmentation*. In a more primitive unsegmented animal, such as a flatworm, the muscles run continuously through the length of the body. A soft-bodied, legless insect larva with its simple motor mechanism can contract its body, bend from side to side, or coil itself in a circle; in short, it can make most of the movements of a segmented worm.

The adult insect with its sclerotized integument is confronted with a new problem in the mechanics of movement. The longitudinal muscles retain their primary points of attachment on the interseg-

27

mental grooves (fig. 9 A, *lmcl*), but the sclerotization cannot conform with the primary segmentation in the soft larval integument. Where the segments preserve their individuality, as in the abdomen, the sclerotization forms in each segment a dorsal plate,

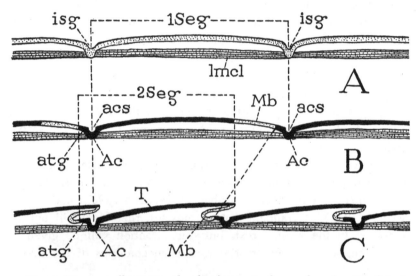

Fig. 9. Diagrams illustrating the development of secondary functional body segmentation in a sclerotized integument from the primary segmentation of a soft-skinned larva.

A, primary dorsal segmentation in the soft larval integument, with longitudinal muscles attached on the intersegmental grooves. B, secondary segmentation in the sclerotized integument; the intersegmental grooves become internal ridges (antecostae) of the segmental plates, and the unsclerotized parts of the preceding plates become the functional intersegmental membranes of the secondary segments. C, the secondary segments drawn together, resulting in infolding of the intersegmental membranes.

For explanation of abbreviations see pages 29–30.

or *tergum*, and a ventral plate, or *sternum*, the two separated on the sides by nonsclerotized areas to allow of dorsoventral expansion and compression. The sclerotization of the dorsum and the venter extends forward in each segment and includes the inflection on which the muscles are attached (B), so that the primary intersegmental fold becomes a solid internal ridge, the *antecosta* (*Ac*), of the segmental plate, marked externally by an *antecostal sulcus* (*acs*). Usually the sclerotization extends a short distance beyond the sulcus as a precostal flange, the *acrotergite* (*atg*) of a dorsal plate or the *acrosternite* of a ventral plate. Posteriorly the sclerotization does not reach the full length of the segment but leaves a flexible membranous area of the integument (*Mb*) between it and the following plate. By this device the body wall of the adult insect is divided into sections movable on each other by the attached muscles, but it is now to be seen that the body sections (B, *2Seg*), which are the functional segments of the sclerotized insect, do not entirely correspond with the primary larval segments (A, *1Seg*) and that the so-called intersegmental membranes (B, *Mb*) are the posterior parts of the primary larval segments. We have here, therefore, a *secondary segmentation* of the body, which allows movement of the functional segments on one another and, by an infolding of the intersegmental membranes (C, *Mb*), permits the segmental annuli to be telescoped in the manner they are usually seen in the abdomen of an adult insect.

The sclerotic pattern of the body, however, is not always so diagrammatically simple as that just described, since various adjustments may be made to accommodate special mechanisms. The segments themselves, furthermore, may be firmly united, as in the thorax, or entirely consolidated, as in the head. Finally, the muscles do not always retain their primitive attachments on the antecostae. Anteriorly in the adult insect they often arise on the postcostal parts of the terga and sterna, and posteriorly they are sometimes attached on the precostal flanges.

Explanation of Abbreviations on Figures 4–9

Ac, antecosta.
acs, antecostal sulcus.
Ap, apodeme.
atg, acrotergite.

BMb, basement membrane.
BW, body wall.

Cl, epidermal cell.
Ct, cuticle.

E, end disc of muscle fiber.
Enct, endocuticle.
Epct, epicuticle.
Exct, exocuticle.

Fbl, muscle fibril, sarcostyle.

Fbr, muscle fiber.

Gng, ganglion.

Ht, heart.

isg, intersegmental groove.

J, light disc of muscle fiber.

$L_1, L_2, L_3,$ leg buds.
lmcl, longitudinal muscles.

M, mesophragma of muscle fiber.
Mb, secondary intersegmental membrane.
Mcl, muscle.

29

N, accessory disc of muscle fiber.
NC, ventral nerve cord.
Nu, nucleus.

Ph, phragma.
Prct, procuticle.

Q, dark disc of muscle fiber.

R, internal skeletal ridge.

s, sulcus.
Sar, sarcoplasm.
Sarl, sarcolemma.
Seg, segment (*1Seg,* primary segment, *2Seg,* secondary segment).

Sp, spiracle (*1Sp,* first spiracle, *2Sp,* second spiracle).
Srs, sarcosomes.

T, tergum.
Tfbl, tonofibrillae.
Tmg, tormogen.
Tra, trachea.
TraSc, tracheal sac.
Trg, trichogen.

W_2, W_3, wing buds.

Z, telophragma of muscle fiber.

III

THE HEAD

THE head of an adult insect is a craniumlike capsule separated from the thorax by a distinct neck. It bears the eyes, the antennae, the labrum, the mouth, and the organs of feeding and contains the brain, the suboesophageal ganglion, and the anterior parts of the food tract. In contrast to the thorax and the abdomen, the adult head gives little suggestion of being a segmented part of the insect, but in its development the head is seen to be a composite structure formed by the union of several primary body segments with a primitive head lobe.

DEVELOPMENT OF THE HEAD

A good example of an embryo in which the head components are distinct is furnished by the young embryo of a locust (fig. 10 A), as illustrated by Roonwal (1936). The head at this stage consists of a large cephalic lobe (*CL*), on which are situated the mouth, which is the orifice of the stomodaeal ingrowth (*Stom*), and rudiments of the antennae (*Ant*). Behind the head lobe are three body segments bearing the rudiments of the future mouth parts (*Md, 1Mx, 2Mx*), followed by the three thoracic segments (*Th*) and an unsegmented abdomen (*Ab*). At a later stage (B), the labrum (*Lm*) appears in front of the mouth, and the abdomen becomes segmented. This type of structure is characteristic of the early embryo in the lower orders of insects, as may be seen in numerous examples illustrated by Johannsen and Butt (1941), representative of Collembola, Thysanura, Odonata, Embiidae, Isoptera, Orthoptera, and Hemiptera, and typically present in a young cockroach embryo (see Hagan, 1951, figs. 124, 125). From these insects it is clear that the mouth-part segments are at first merely a part of the body; it is only at a later stage that they combine with the cephalic lobe to form the definitive head. The higher insects, however, eliminate this early stage of head development. In the

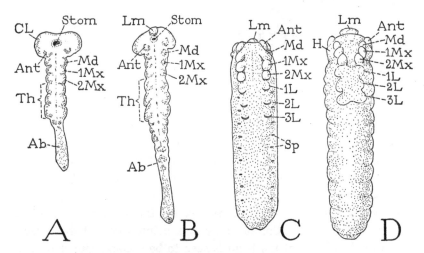

Fig. 10. Development of the insect head and mouth parts.
A, 53-hour embryo of a grasshopper (*Locusta*) with distinct cephalic lobe (*CL*) and mouth-part segments a part of the body (from Roonwal, 1936). B, 59-hour embryo of same (from Roonwal). C, 48–50-hour embryo of honey bee, ventral, cephalic lobe, and mouth-part segments united (from Nelson, 1915). D, 58–60-hour embryo of same (from Nelson).
For explanation of abbreviations see pages 42–43.

honey bee the embryo at the 50-hour stage of growth, as shown by Nelson (1915), is of uniform width throughout (fig. 10 C), there being no specific head lobe, but here again the mouth-part segments are in no way differentiated from the thoracic region. At a later stage (D), the mouth-part rudiments are crowded forward, and the definitive head region (*H*) bearing the labrum, the antennae, and the mouth parts is separated by a slight cervical constriction from the thorax.

Though the cephalic lobe of the embryo gives no external evidence of its ever having been itself a segmented region of the trunk, most students of arthropod "head segmentation," basing their conclusions on a study of primitive coelomic sacs, nerve distribution, and even the external grooves of the insect head, contend that the cephalic lobe itself must have originally been formed by the union of several primitive segments representing the anterior segments of the annelid worms. There are others, however, who interpret the embryonic cephalic lobe of the arthropods as an unsegmented prostomium. If there ever was a stage in the evolution of the arthropods when the cephalic lobe was segmented, the animals lived in remote Pre-Cambrian times and probably will never be brought to the witness stand.

Ignorance of prehistoric facts will not interfere with an under-standing of the structure of the head of the modern adult honey bee. It must be understood, however, that the bee embryo (fig. 1 E) develops directly into the larva, in which the head and feeding organs (fig. 18) are quite different from these parts in the adult, because they are adapted specifically to serve the larva shut up in a cell of the comb, where it is fed by the nurses. At the end of the larval period the head and mouth parts are then wholly reconstructed in the pupa to conform with the needs of the free-living adults; in their reconstruction they revert to the fundamental plan of insect structure, but on this basis they build up a highly specialized type of structure characteristic of adult bees.

THE EXTERNAL STRUCTURE OF THE BEE'S HEAD

The shape and general external appearance of the head of a worker bee are shown from before and behind at A and B of figure 11, but structural details are better seen on the outline drawings of figure 12. In facial view (fig. 11 A) the head is triangular with the apex below, from which are appended the mouth parts; the rounded lateral angles are capped by the great *compound eyes* (E). In a side view (fig. 12 C) the head is seen to be much flattened anteroposteriorly, the face being convex, while the rear surface (fig. 11 B) is somewhat concave to fit snugly on the anterior end of the thorax. Three simple eyes, or *ocelli* (A, O), are situated on the top of the head, with the median ocellus in front of the other two. The *antennae* (Ant) arise close together from the face between the lower halves of the compound eyes; their long basal segments are inserted into small circular membranous sockets of the head wall. From the lower edge of the face hangs a broad, movable flap (Lm), which is the *labrum*. The bee's labrum has a single pair of muscles arising on the facial wall of the head between the antennal bases, and attached on the labrum by long tendons arising laterally

33

from the base of the inner wall of the hollow labrum. Behind the labrum is the functional mouth of the bee (A, *Mth″*).

The cranial walls are not marked by grooves in the bee to the extent that they are in some of the lower insects, but the surface regions of the head are given names for descriptive purposes. The

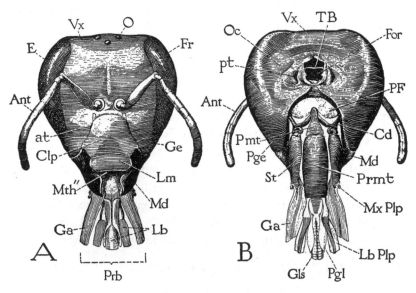

Fig. 11. Head of a worker bee. A, anterior; B, posterior. For explanation of abbreviations see pages 42–43.

top of the head is the *vertex* (figs. 11 A, 12 A, *Vx*); the upper part of the face between the compound eyes may be termed the *frons* (*Fr*); the "cheeks" below the eyes are the *genae* (*Ge*). The only facial area of the bee's head limited by a distinct groove is the large shieldlike plate (*Clp*) below the antennae, from which the labrum is suspended. This plate is the *clypeus;* the limiting groove (fig. 12 A, *es*) is the *epistomal sulcus.* In most insects the clypeus projects as a free lobe below the frons, but in the bee and some other Hymenoptera it appears to have been extended upward between the genae into the area of the frons. DuPorte and Bigelow (1953) and Bigelow (1954), however, from a comparative study of the hymenopterous head contend that the genae have been lengthened downward and have thus enclosed the clypeus. The clypeus gives attachment to important muscles of the sucking organ within the head (fig. 17 B, *dlcb*), and it may be that its position is correlated with the extent of these muscles. A pair of lateral pits in the upper part of the epistomal sulcus (fig. 12 A, *at*) mark the anterior roots of the endoskeletal tentorium of the head (D).

The posterior wall of the cranium (figs. 11 B, 12 B) is perforated

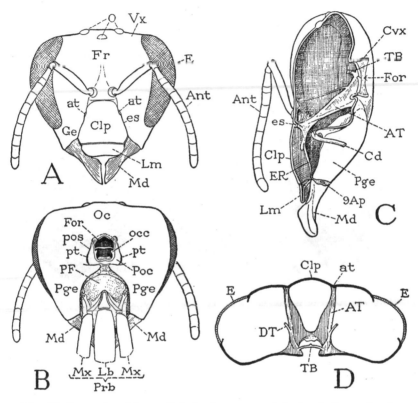

Fig. 12. External and internal skeletal structure of the head of a worker bee. A, anterior view of head. B, posterior. C, right half of head, mesal view, showing inner skeleton. D, horizontal section of head above tentorium.

For explanation of abbreviations see pages 42–43.

by a pentagonal hole (*For*), the *occipital foramen*, analogous to the foramen magnum of a vertebrate skull, by means of which the head cavity communicates through the membranous neck with the cavity of the thorax. The foramen gives passage to the oesophagus, the dorsal blood vessel, air tubes, nerves, and the salivary ducts that extend between the head and the thorax. The area of the head wall above and at the sides of the foramen is termed the *occiput* (*Oc*); the lower lateral areas are the *postgenae* (*Pge*). Surrounding the

35

foramen is a *postoccipital sulcus* (fig. 12 B, *pos*), which sets off a narrow *postocciput* (*Poc*) immediately enclosing the foramen. In the lateral parts of the postoccipital sulcus are two slitlike depressions (*pt*) which are the posterior roots of the tentorium. Below the foramen the sclerotic wall of the cranium is deeply excavated between the postgenae by an inverted U-shaped depression (*PF*) with a membranous floor, from which the basal parts of the proboscis (*Prb*) are suspended, and which is therefore known as the *fossa of the proboscis*. The fossa is in part separated from the occipital foramen by convergent lobes of the postgenae.

The head of the queen (fig. 13 B) differs slightly from that of the

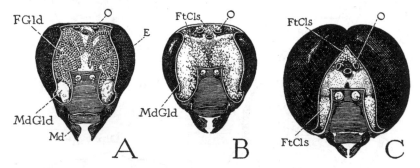

Fig. 13. Head of a worker (A), queen (B), and drone (C), with most of facial area removed.

For explanation of abbreviations see pages 42–43.

worker (A) in being more rounded on the sides and a little wider in proportion to its length. The head of the drone (C) is much larger than that of either of the female forms and is nearly circular in facial view because of the greatly enlarged compound eyes, which are contiguous on the vertex. The ocelli of the drone (*O*) appear to have been displaced downward on the face by the union of the compound eyes above them.

THE INTERNAL HEAD SKELETON

The internal skeleton of the head consists principally of the structure known as the *tentorium,* but there is a strong *epistomal ridge* (fig. 12 C, *ER*) along the line of the epistomal sulcus enclosing the clypeus and in the drone a deep mid-cranial inflection of the vertex between the upper ends of the compound eyes (fig. 14 B, *mcr*). The tentorium of the bee is in no sense "tent-shaped" as the name would imply. It consists principally of two thick, strongly sclerotized bars (fig. 12 D, *AT*), extending from the anterior tentorial pits (*at*) in the epistomal sulcus posteriorly and upward (C) to the posterior pits in the postoccipital sulcus, and includes a nar-

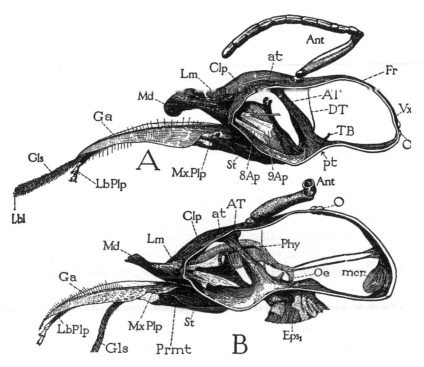

Fig. 14. Longitudinal section of head of a worker (A) and a drone (B), cut on left side of median plane.

For explanation of abbreviations see pages 42–43.

row arched rod (D, *TB*) forming a bridge between the posterior ends of the bars. Shortly before the bridge each longitudinal bar gives off dorsally a slender branch (*DT*), which extends forward (C) but disappears before it reaches the facial wall of the head. The oesophagus and the dorsal blood vessel go over the tentorial bridge in traversing the occipital foramen; the nerve cords and the salivary ducts go beneath it.

The insect tentorium is formed by paired anterior and posterior arms that grow in from the anterior and posterior tentorial pits, respectively. The posterior arms join to form a transverse bridge through the back of the head; and in most insects the anterior arms unite posteriorly with the bridge, but there is much variation in the relative development of the parts. In the bee the anterior arms (figs. 12 C, D; 14 A, B, *AT*) are greatly enlarged, and the posterior bridge is reduced to a narrow bar (*TB*). The dorsal arms (*DT*) are secondary outgrowths of the anterior arms; although in most insects they unite with the dorsal or anterior head wall, they are not formed as ingrowths of the latter. In some insects the tentorial arms are united in a wide central plate, in which case the tentorium might be likened to a canopy suspended by four stays, whence probably the name "tentorium" was conceived in the imagination of some early entomologist. The tentorium serves as a brace between the head walls and also gives attachment to muscles of the antennae and the mouth parts.

THE MUSCLES OF THE HEAD

The head is attached to the thorax by a relatively narrow, membranous neck, or *cervix*. In the bee there are no cervical sclerites; the head is supported directly on the body by two processes of the anterior ends of the episternal plates of the thorax (fig. 31 B, *e*). These processes articulate with corresponding *occipital condyles* on the lateral margins of the postocciput of the head (fig. 12 B, *occ*). The head muscles come from the prothorax and are attached on the margin of the postocciput of the cranium (*Poc*). According to Nelson (1915), the posterior arms of the tentorium arise in the embryo between the bases of the first and second maxillae. Since in the adult bee the posterior tentorial pits (*pt*) lie in the postoccipital sulcus (*pos*), this groove would appear to be the dividing line between the maxillary and labial segments. If so, the narrow postoccipital flange, supported directly on the articular arms of the prothoracic episterna, is the only sclerotic remnant of the labial segment. In most insects the head muscles are attached on the internal ridge of the postoccipital sulcus.

The head of the bee is movable by five pairs of muscles from the thorax attached on the margins of the occipital foramen, four pairs being attached dorsally above the occipital condyles and one pair ventrally. Of the dorsal muscles, two pairs (fig. 15 A, *40, 41*) arise on the phragma (*1Ph*) between the pronotum and the mesonotum; the second pair of these muscles (*41*) are attached in the apical notch of the foramen, the other pair (*40*) just at the sides

38

of the notch. The third dorsal pair are large, three-branched muscles arising by broad bases on the prothoracic episterna (D, 42); the fibers of each muscle converge to a strong tendon attached dorso-laterally on the foraminal margin (B, 42a). The fourth dorsal

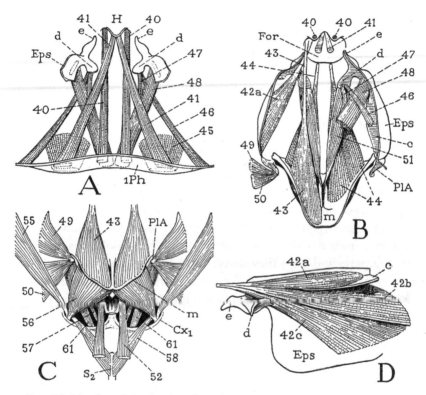

Fig. 15. Muscles of the head and prothorax.

A, dorsal muscles in prothorax, ventral. B, ventral and lateral muscles, dorsal. C, muscles of endosternum and coxa of prothorax, posterior. D, branches of head muscle *42* arising on proepisternum.

c, horizontal apodemal ridge of episternum; *d,* cervical apodeme of episternum; *e,* occipital process of episternum; *m,* supraneural bridge of endosternum. Muscles are explained in text. For explanation of abbreviations see pages 42–43.

muscles arise on the upper surface of the prothoracic endosternum (B, C, 43) and are attached by long tendons (B) just laterad of the episternal muscles. The muscles of the single ventral pair are large two-branched muscles (B, 44) arising on the prothoracic endosternum beneath 43 and inserted by tendons on the lower margin of the head foramen. It would appear that the head muscles serve principally to turn the head up or down on the axis between the occipital condyles; but if the muscles of opposites sides act antagonistically, these same muscles might be supposed to turn the head from side to side as far as the flexibility of the supporting episternal plates will permit. The episterna themselves have strong muscles from the prothoracic endosternum (fig. 15 B, *51*), and others from the phragma (A, *46, 47, 48*), all attached at the bases of the head-supporting processes, and these muscles may play some part in the movement of the head. Evidently the bee cannot revolve its head on two fixed points of articulation unless the episternal processes themselves are movable.

THE ANTENNAE

The antennae may appropriately be described in connection with the head because, as we have seen, they are primitive cephalic appendages developed on the head lobe of the embryo (fig. 10 A, B). In their structure the antennae give no evidence of having been evolved from a pair of appendages serially homologous with the mouth parts and legs; they derive their innervation from the supra-oesophageal brain. Even in the ancient trilobites the antennae are long tapering filaments resembling the antennae of a modern cockroach.

The antennae of the bee are formed as small protuberances on the head of the embryo (fig. 10 C, D, *Ant*), but, according to Nelson (1915, 1924), prior to hatching they become reduced to mere thickenings of the epidermis, as do also the legs. In the larva the antennal cells renew their development, but the larval antennae grow within pockets of the head epidermis beneath the cuticle, their position being marked externally only by the slightly raised discs seen on the face of the larva. The antennae complete their development during the propupal stages (fig. 2 E, G, *Ant.*, H) and are exposed on the pupa (I) with the shedding of the last larval cuticle. Though the antennal rudiments appear on the embryo behind the mouth, the organs subsequently migrate forward and upward to their definitive facial position above the clypeus (fig. 11 A).

Each antenna of the adult bee (fig. 16 A) is divided by an elbow joint into a rigid basal stalk, or *scape* (*Scp*), and a long, flexible distal part subdivided into 11 small sections in the female (A) and

12 in the male (G). The basal section of the latter is named the *pedicel* (A, *Pdc*) and the part beyond it the *flagellum* (*Fl*). The pedicel contains a special sense organ, and in many insects it is more distinctly differentiated from the flagellum than in the bee.

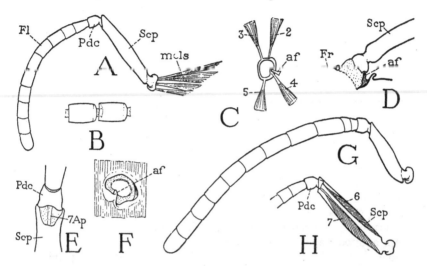

Fig. 16. Structural details of the antenna.
A, right antenna of worker and its muscles, mesal. B, subsegments of flagellum pulled apart to show connection. C, muscle insertions on base of antenna. D, base of right antenna and antennal socket of head, posterior. E, joint between scape and pedicel, ventral. F, rim of right antennal socket, with antennifer. G, antenna of drone. H, proximal part of worker antenna, showing muscles of pedicel arising in scape.
For explanation of abbreviations see pages 42–43.

The flagellar subdivisions are not true segments, since they are not articulated upon each other or provided with muscles. The rounded bases of each section (B) fits into a distal depression of the one proximal to it, and the two are connected by a narrow membranous neck. The joint between the pedicel and the scape, on the other hand, is a dicondylic hinge (E) allowing the flagellum to be turned up or down on a definite transverse axis on the end of the scape. The flagellum is movable on the scape by two muscles arising in the latter (H, *6, 7*), which are attached on the base of the pedicel. These are the only muscles within the antenna of the bee.

The antennal scape is attached to the head by a basal knob set into a membranous socket of the cranial wall (fig. 16 D), and the wall is strengthened by a marginal thickening (F). The scape is pivoted on a single articular process, the *antennifer* (*af*), arising ventrolaterally from the rim of the socket. The antenna as a whole is thus freely movable in all directions except as it is limited by the socket membrane. To provide for its movement, four muscles arising on the anterior tentorial arm of the same side are inserted on the base of the scape (A, *mcls*), two attached above the level of the antennifer and two below it (C).

The antennae are the seat of numerous sense organs, particularly those of the tactile and olfactory senses, but the antennal sense organs will be described in the chapter on the nervous system and the sense organs in general.

Explanation of Abbreviations on Figures 10–16

Ab, abdomen.

af, antennifer.

Ant, antenna.

7Ap, apodemal tendon of ventral muscle of antennal pedicel.

8Ap, abductor apodeme of mandible.

9Ap, adductor apodeme of mandible, or its attachment point.

at, anterior tentorial pit.

AT, anterior tentorial arm.

Cd, cardo.

CL, cephalic lobe of embryo.

Clp, clypeus.

Cvx, cervix, neck.

Cx, coxa.

DT, dorsal tentorial arm.

E, compound eye.

Eps, episternum.

Eps₁, episternum of prothorax.

ER, epistomal ridge.

es, epistomal sulcus.

FGld, food gland.

Fl, flagellum of antenna.

For, occipital foramen.

Fr, frons.

FtCls, fat cells.

Ga, galea.

Ge, gena.

Gls, glossa.

H, head.

1L, 2L, 3L, legs or leg rudiments.

Lb, labium.

Lbl, labellum.

LbPlp, labial palpus.

Lm, labrum.

mcls, basal muscles of antenna.

mcr, midcranial ridge.

Md, mandible.

MdGld, mandibular gland.

Mx, maxilla.

1Mx, first maxilla.

2Mx, second maxilla.

MxPlp, maxillary palpus.

O, ocelli.

Oc, occiput.

occ, occipital condyle.

Oe, oesophagus.

Pdc, pedicel of antenna.

PF, proboscis fossa.

Pge, postgena.

Pgl, paraglossa.

1Ph, first thoracic phragma.

Phy, pharynx.
PlA, pleural arm.
Pmt, postmentum.
Poc, postocciput.
pos, postoccipital sulcus.
Prb, proboscis.
Prmt, prementum.
pt, posterior tentorial pit.

S₂, mesothoracic sternum.

Scp, scape of antenna.
Sp, spiracle.
St, stipes.
Stom, stomodaeum.

TB, tentorial bridge.
Th, thorax.

Vx, vertex.

✦ IV ✦

THE ORGANS OF FEEDING

THE organs of an insect directly concerned with the intake of food are the *mandibles,* the *maxillae,* and the *labium.* These organs are postoral segmental appendages equivalent in origin to the legs of the thorax, the single labium being formed by the union of a pair of second maxillae (fig. 22 D). An important part of the ingestion apparatus, however, is the space between the mouth parts covered by the clypeus and labrum, which thus becomes a *preoral cavity* into which the food is received before it is passed into the mouth and into which the glands of the head open. Within the preoral cavity is a median postoral lobe of the ventral head wall known as the *hypopharynx.* This intergnathal preoral food cavity, being entirely outside the mouth, is neither a "buccal cavity" nor a "pharynx," though it has often been so called.

The feeding apparatus of the honey bee is made up of the same anatomical parts as is that of a generalized insect, including the preoral cavity and the hypopharynx, the mandibles, the maxillae, and the labium, but the maxillae and the labium are integrated to form a *proboscis* for the intake of liquids. All these parts, however, are greatly modified structurally, and new mechanisms have been evolved in them as special adaptations to the needs of the bee. An interpretation of the parts of the feeding apparatus of the bee in terms of fundamental insect anatomy, therefore, is in many respects difficult, unless we closely compare the organs of the bee with the corresponding organs in a more generalized insect. It must not be forgotten that the bee is an insect, and, with all its special modifications that enable it to live the life of a bee, it cannot get away from its basic ancestral anatomy. In studying the feeding apparatus of

44

the bee, we must understand first the structures involved in the preoral cavity, including the sucking pump, the hypopharynx, and the openings of the head glands, since the function of the external organs, particularly of the proboscis, depends on these concealed parts.

THE PREORAL CAVITY AND THE HYPOPHARYNX

Since the structures within the preoral cavity are highly special ized in the bee, they will be more easily understood if we first examine the corresponding parts in some generalized insect, and for this purpose the cockroach will serve very well. A vertical section through the head of a cockroach just to one side of the median plane (fig. 17 A) discloses the preoral cavity between the clypeus and labrum (*Clp, Lm*) in front and the prementum (*Prmt*) of the labium behind. Normally the cavity is shut in on the sides by the mandibles and maxillae. The so-called preoral "cavity" is thus seen to be merely an external space enclosed by the mouth parts. The true mouth of the insect (*Mth'*) lies at the inner end of the preoral cavity and leads into the pharynx (*Phy*), which is the first part of the stomodaeum. The distal opening into the preoral cavity between the labrum and the labium is the functional mouth (*Mth''*), since it is the actual food-intake orifice. From the oblique inner wall of the preoral cavity behind the mouth hangs the tonguelike hypopharynx (*Hphy*). Behind the hypopharynx at the base of the labial prementum is the orifice of the salivary duct (*slO*).

The mandibles close upon each other between the labrum and the hypopharynx. Proximal to the mandibles is a pocket of the preoral cavity (fig. 17 A, *Cb*) between the inner wall of the clypeus (*Clp*) and the base of the hypopharynx, into which the food is received from the jaws to be passed on into the mouth at its inner end. This preoral food-receiving pocket is the *cibarium*. The clypeal wall of the pocket is covered with transverse muscles (not shown in the figure), and on it are attached large muscles (*dlcb*) from the cranial wall of the clypeus. The cibarium, therefore, can be compressed and dilated, and very probably it serves the cockroach for the ingestion of liquids; in liquid-feeding insects the cibarium is developed into a highly efficient sucking pump.

The long, thick hypopharynx of the cockroach consists of a free tonguelike lobe (fig. 17 A, *Hphy*) and of a basal part that slopes

45

anteriorly to the mouth. In each side of the base is a small sclerite, the two of which are connected by a narrow sclerotic bridge across the anterior hypopharyngeal surface, and each is continued upward in a slender arm (y) that goes through the mouth angle of the same side and gives attachment to two muscles, one muscle arising on the frons, the other on the ventrolateral area of the cranium. This structure is the *hypopharyngeal suspensorium* (*HS*). The oral arms going through the mouth angles mark the dividing line between the preoral cibarium (*Cb*) and the postoral pharynx (*Phy*). The posterior wall of the hypopharynx ends at the orifice of the salivary duct (*slO*) at the base of the labial prementum. There is thus a posterior pocket of the preoral cavity between the hypopharynx and the labium. This pocket is the *salivarium* (*Slv*), so named because it receives the secretion of the salivary glands, which open by a median duct (*slDct*) into its inner end.

The pharynx of the insect, properly so called, is a differentiation of the stomodaeal section of the alimentary canal posterior to the mouth. In the cockroach it is a somewhat enlarged region of the stomodaeum (fig. 17 A, *Phy*). In liquid-feeding insects the pharynx may become a second, postoral organ of suction, but it is always to be distinguished from the cibarium by the fact that its dilator muscles are separated from those of the cibarium by the brain connectives of the frontal ganglion (*frGng*). Unfortunately the earlier entomologists gave the name "pharynx" to the cibarium and thereby left us a heritage of misapplied terms. The inner wall of the labrum and clypeus, for example, is still known as the "epipharyngeal" surface, and we have only the term "hypopharynx" for the intergnathal lobe so named, though the latter is developed from the ventral wall of the head entirely outside the pharynx.

If we now compare a section of the bee's head (fig. 17 B) with that of the cockroach (A), we see that in general structure the two are similar; but there are several features in the bee that at first sight appear to be quite different. A strict comparison of the parts in the bee with those of the cockroach, however, will show that the two insects are fundamentally alike and that we may correctly interpret the specialized bee structure in terms of generalized insect anatomy. In some respects, however, the results will not conform with interpretations and names that have become current in anatomical bee literature.

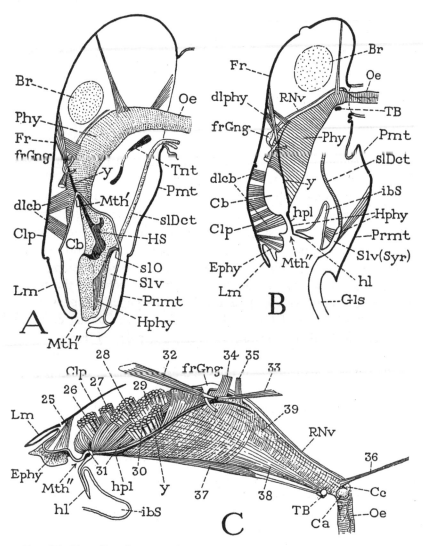

Fig. 17. The cibarial pump, the pharynx, and the hypopharynx.

A, vertical section of head of a cockroach, showing preoral cibarium (*Cb*) separated from postoral pharynx (*Phy*) by arms (*y*) of hypopharyngeal suspensorium. B, corresponding section of head of a worker honey bee. C, the cibario-pharyngeal sac of the bee.

For explanation of abbreviations see page 79.

The most conspicuous organ in the head of the bee is a large muscular sac that extends upward from the functional mouth (fig. 17 B, *Mth″*) and narrows into the oesophagus (*Oe*), which turns backward through the neck. This sac has commonly been called the "pharynx" of the bee; but the error of identifying the entire sac with the pharynx should be at once evident on observing that its walls are traversed on each side by a long slender rod (B, C, *y*) that gives attachment dorsally to two opposing muscles (C, 32, 33) and that the pair of rods arise from a plate (*hpl*) on the floor of the functional mouth. Comparison with the cockroach (A) then leaves no doubt that these rods and the suboral plate represent the hypopharyngeal suspensorium. The upper ends of the rods, therefore, divide the head sac of the bee into a dorsal pharyngeal section (B, *Phy*), belonging to the stomodaeum of the alimentary canal, and a lower cibarial section (*Cb*), which is a part of the preoral cavity. The true mouth of the bee, as in the cockroach, is the opening from the cibarial region of the sac into the pharynx between the upper ends of the oral arms (*y*) of the hypopharyngeal suspensorium.

The cibarial part of the head sac is the active sucking organ of the bee, as in other sucking insects. Five pairs of large dilator muscles from the clypeus (fig. 17 B, *dlcb*, C, 26–30) are inserted on its anterior wall between thick bands of transverse compressor muscles (C, 31). As in other sucking insects the cibarial pump has been evolved from an open food pocket by a lateral union of the inner clypeal wall with the hypopharynx along the oral rods. The base of the hypopharynx thus forms the floor, or posterior wall, of the pump, and in some sucking insects it becomes a strongly sclerotized trough or basin. The pharyngeal section of the head sac in the bee (C) is ensheathed in layers of circular and longitudinal muscle fibers and evidently must be strongly contractile, probably serving to drive the food received from the cibarial pump back into the oesophagus.

Inasmuch as the floor, or posterior wall, of the cibarium is the basal part of the anterior wall of the hypopharynx (fig. 17 A), the suboral plate of the bee (B, *hpl*) belongs to the hypopharynx and is not a "pharyngeal" plate, as it has commonly been called. From the margin of the hypopharyngeal plate there depends a small, flat, bifid, triangular lobe (figs. 17 B, C, 21 A, *hl*) in the position of the hypopharyngeal lobe of the cockroach, but the undersurface of the hypopharynx in the cockroach extends to the opening of the salivary duct. In the bee, therefore, the entire surface (fig. 17 B, Hphy) from the suboral plate (*hpl*) to the salivary orifice is derived from the hypopharynx. The hypopharyngeal wall behind the suboral lobe is first inflected as an open pouch (*ibS*) and is then extended and united with the margins of the prementum of the labium, thus converting the salivarium (*Slv*) into a closed pocket, or salivary exit passage, opening on the distal end of the prementum (fig. 26 B, *slvO*). It is readily seen now that if all this infolded hypopharyngeal surface of the bee were everted, the bee would have a typical hypopharynx corresponding with that of the cockroach.

The union of the hypopharynx with the labium is best seen in the larva of the bee. From the larval head (fig. 18 A) there projects

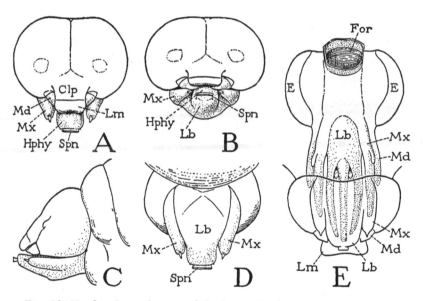

Fig. 18. Head and mouth parts of the honey-bee larva and pupa.
A, head of larva, anterior. B, same, ventroanterior. C, same, lateral. D, same, posterior. E, head of propupa of worker, posterior, and outline of larval head, showing origin of pupal mouth parts in corresponding parts of larva.
For explanation of abbreviations see page 79.

below the mouth and between the ends of the maxillae (*Mx*) a thick median lobe, on which is situated the opening of the ducts of the silk glands (*Spn*). Since the larval silk glands represent the salivary glands of adult insects, their normal opening is between

49

the base of the hypopharynx and the base of the labium. From this fact we know that the hypopharynx of the bee larva (B, *Hphy*) must be united with the labium (*Lb*) in the median lobe of the mouth parts, leaving open through the lobe only the lumen of the salivarium, which thus becomes the exit passage for the silk glands and is not itself a part of the silk duct. Nelson (1924) regarded the median lobe of the larval mouth parts as the labium only and said (1915, p. 105) that the silk duct "opens at the tip of the labium," but the correct interpretation of this lobe in the bee larva as a "labium-hypopharynx" is given by Dobrovsky (1951), and the lobe is so described by Short (1952) in *Bombus* and *Vespula*. Unfortunately Dobrovsky calls the cibarium the "pharynx," though he clearly shows (fig. 5) its dilator muscles arising on the clypeus anterior to the frontal ganglion.

The infolding of the hypopharyngeal wall behind the suboral lobe (fig. 17 B, *ibS*) forms in some Hymenoptera a definite pocket termed the *infrabuccal sac,* or *gnathal pouch.* The sac is said by

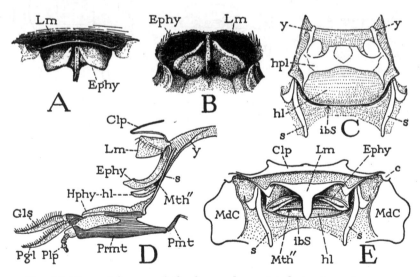

Fig. 19. The epipharynx of the bee and associated structures in wasps.

A, labrum and epipharynx of honey bee, ventral. B, same, posterior. C, hypopharyngeal plate and opening of infrabuccal sac of *Vespula maculata*. D, labrum, epipharynx, hypopharynx, and labium of *Scolia nobilitata*, lateral. E, mouth region of *Vespula maculata*.

c, clypeal articulation of mandible; *s,* suspensory rod of labium from base of labrum; *y,* oral arm of hypopharyngeal suspensorium. For explanation of abbreviations see page 79.

Duncan (1939) to be formed in the wasp, *Vespula,* as an invagination of the hypopharynx; its mouth is ordinarily closed in the wasp
to a narrow transverse slit (fig. 19 E, *ibS*) beneath the edge of the
suboral lobe (*hl*) of the hypopharynx. In both the bee and the wasp
a pair of suspensory rods of the labium (figs. 19 C, E, 27 D, *s*) extend upward from the base of the prementum past the sides of the
infrabuccal sac to the lateral angles of the labrum.

The infrabuccal pocket of the bee appears to be merely an infolding of the hypopharyngeal surface incidental to the retraction
of the labium, and so far as known has no particular function. In
some other Hymenoptera, however, it serves as a trash receptacle.
In the ants, Wheeler (1923) says, solid bits of food and dirt particles and fungus spores cleaned off the body by the ant are lodged
in the sac, where they are compressed to a pellet, which, after
juices that may be present are sucked out, is ejected. In the wasp,
Vespula, Duncan (1939) observes that the pouch "serves as a receptacle for dirt, detritus, and solid matter strained out of the food"
and "probably receives also dirt and debris which the wasp removes
from its body during cleaning operations."

THE HYPOPHARYNGEAL GLANDS

The food material produced by worker bees, known as royal
jelly, or bee milk, which is fed to the larvae, queen, and drones, is
secreted in a pair of long glands distributed in many loops and coils
in the sides of the head (fig. 20, *FGld*). The ducts open on the
suboral plate of the hypopharynx (fig. 21 B). The food glands of
the bee, therefore, are *hypopharyngeal glands* and not "pharyngeal,"
glands as they have commonly been called. Their structure has been
described by various writers, including Schiemenz (1883), Heselhaus (1922), Soudek (1927), Bugnion (1928), Kratky (1931), and
Beams and King (1933).

Each gland consists of numerous small oval cellular bodies attached by short necks to an axial duct (fig. 21 B, *FGld, G*). Soudek
(1927) says there are as many as 550 lobules on each duct. When
fully stretched out a single gland is found to be much longer than
the entire body of the bee. The ducts discharge separately through
two small terminal pouches (B) opening on the distal angles of
the hypopharyngeal plate (A, *o*).

The lobules of the glands are solid cellular bodies, but each cell

51

is individually connected with the axial duct by a fine tubule. The presence of the tubules may be demonstrated by treating a part of a gland with weak caustic, which dissolves the cytoplasm of the cells and exposes the groups of tubules. The structure of the gland

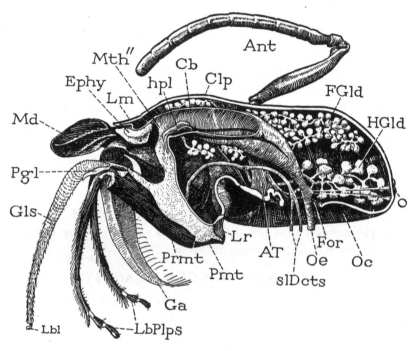

Fig. 20. Right half of head of worker bee with mouth parts attached, cut a little to left of median plane, muscles and nerve tissue omitted.

For explanation of abbreviations see page 79.

cells and their inclusions are fully described by Soudek, who says the cytoplasm is dense, finely granulated, and much vacuolated; he notes also a structural difference between the intracellular and extracellular parts of the ductules. According to Beams and King (1933), the parts of the ductules within the cells are devoid of a cuticular lining; they take a sinuous course, coil about the nucleus, and end blindly in the cytoplasm (fig. 21 H). On emerging from the body of the cell, however, the ductules acquire a cuticular intima and run parallel through the neck of the lobule to open individually into the main duct.

Inasmuch as the food material from the glands is discharged

directly on the suboral plate of the hypopharynx, it would seem that
it must run down the biblike lobe depending from the plate and
accumulate on the base of the labium, where it becomes accessible
to other adult bees. When the labium is retracted (fig. 17 B), the
walls of the infrabuccal sac (ibS) are deeply inflected behind the
hypopharyngeal lobe (hl), and if the labium is now swung forward,

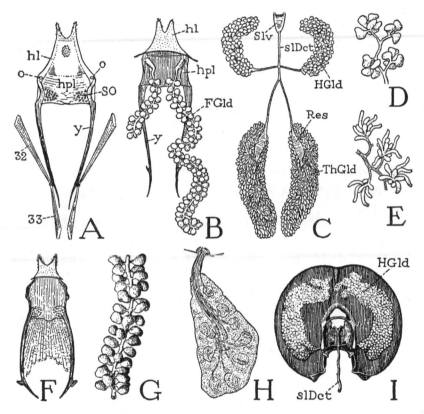

Fig. 21. Glands of the head and thorax.
A, hypopharyngeal plate of worker, anterior, showing apertures (o) of food
glands. B, same, posterior, with proximal parts of food glands. C, labial salivary
gland system of worker, including head glands ($HGld$) and thoracic glands
($ThGld$). D, detail of head gland. E, detail of thoracic gland. F, hypopharyngeal plate of drone. G, detail of food gland. H, section of a food-gland
lobule (from Beams and King, 1933). I, posterior wall of head of worker,
anterior, showing position of head salivary glands.

o, orifice of food-gland duct; y, oral arm of hypopharyngeal suspensorium,
with protractor (32) and retractor (33) muscles. For explanation of abbreviations see page 79.

the lobe gives a direct passage for the food to the labial base. The antagonistic muscles attached on the arms of the hypopharyngeal plate (fig. 21 A, 32, 33) evidently serve to protract and retract the plate, and the movements of the plate perhaps help to dislodge the food upon the labium. This same hypopharyngeal apparatus, however, is present in the queen and the drone (F) and, in some form, is a feature of adult insects in general; hence whatever function it may have in the worker bee is merely the adaptation of a common structure to a particular use.

In the feeding of the larvae the gland food is discharged from between the mandibles into the brood cells. Bugnion (1928) says the royal jelly first accumulates in the crop (*jabot*) and by this organ is regurgitated into the cells. Although nectar and honey are regurgitated from the crop (honey stomach), there is no evidence that the royal jelly is first swallowed into the honey stomach and then expelled from the latter.

The development and physiological activity of the food glands vary with the work in which the bee is engaged, the glands being fully functional when the worker is serving in the hive as a nurse bee feeding the larvae and the queen. According to Soudek (1927), the gland cells are empty or atrophic in bees collecting pollen and nectar; though a high percentage of bees with full glands were taken on the hive entrance, these were probably young bees making orientation flights. A pollen diet appears to be necessary for the full development of the glands. The information at present є ᵃail-able on the physiology of the food glands, their development rₑₗa-tive to the activities of the bee, the nature of their secretion and its probable enzymes is reviewed by Ribbands (1953, pp. 55–59, 245–254).

Glands of the hypopharynx are not generally present in insects other than Hymenoptera. They are particularly large in the worker honey bee but have not been observed in either the queen or the drone. The openings of the ducts on the hypopharyngeal plate, however, are reported to be sometimes present in the queen, and Heselhaus (1922) records finding in one drone specimen not only the apertures but also the terminal pouches of the ducts. Possibly, therefore, the glands themselves were once present in each of the three castes of the honey bee.

Hypopharyngeal glands similar to those of *Apis*, except that the secreting cells occur singly on the ducts, are present in other social bees and occur in both the females and the males. The glands of *Bombus* and *Psithyrus* are fully described by Palm (1949), who says their secretion is acid, rich in protein, and contains invertase and amylase. In these genera, therefore, the gland secretion is evidently salivary in function, since it is not fed to the larvae. Among the solitary bees, as shown by Heselhaus (1922) and Bugnion (1928), there are corresponding hypopharyngeal glands in the females, but in these bees the glands are masses of cells with short individual ducts opening directly on the hypopharyngeal plate. Similar glands are present in the wasps *Eucera* and *Vespa*. In *Andrena praecox* Heselhaus reports the glands are present also in the male bee. In the ants, according to Bugnion, the hypopharyngeal glands are composed of single cells with ductules opening, as in the social bees, into a single main duct on each side.

Groups of cells sometimes observed on the undersurface of the oral plate of the hypopharynx have been described as "sublingual glands." Heselhaus (1922, p. 389), however, says these cells have no ducts, and he regards them as weakly vacuolated fat cells.

THE EPIPHARYNX

A special feature of the bee and of various other Hymenoptera is a lobe projecting behind the labrum from the inner wall of the clypeal region of the head (fig. 17 B, C, *Ephy*). This lobe is known as the *epipharynx*. In the bee the epipharynx is a soft, padlike structure, triangular in shape as seen from behind (fig. 19 B) with a thin median lobe depending like a keel from its posterior surface (A). On each side of the keel is an oval elevation covered with small sense hairs (B). The epipharynx is retractile by a muscle from the outer clypeal wall (fig. 17 C, 25) and in the bee serves mechanically, together with the lacinial lobes of the maxillae, to close the food channel on the base of the proboscis (fig. 24 C). In the wasp, *Vespula*, the epipharynx (fig. 19 E, *Ephy*) fits closely against the suboral lobe (*hl*) of the hypopharynx and closes the functional mouth to a narrow slit (*Mth″*). In *Scolia nobilitata* (D), however, both the epipharynx and the suboral lobe of the hypopharynx are extended in long protruding lips with the mouth opening between them.

THE MANDIBLES AND THEIR GLANDS

The mandible of an insect (fig. 22 B) represents the extreme reduction of a leg (A) in its conversion into an organ of feeding. In many Crustacea the telopodite of the leg (A, *Tlpd*) is retained as a small palpus (B, *Plp*) on the mandible, but the insect mandible is simply the coxa (*Cx*) of a limb from which the telopodite has been eliminated and the distal end variously developed according

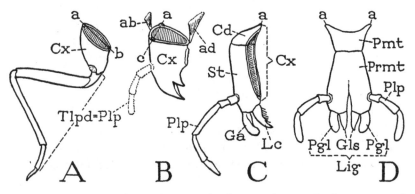

Fig. 22. Diagrams showing structural relations of generalized insect mouth parts to a leg.

A, a typical insect leg. B, a mandible and its usual mechanism. C, a generalized maxilla. D, a simple labium, suggestive of two maxillae united.

a, primary lateral articulation of coxa; *b*, primary mesal articulation of coxa; *c*, secondary anterior articulation of mandible. For explanation of abbreviations see page 79.

to the feeding habits of the insect, or for whatever other purposes the mandibles may be used. The coxa of a primitive leg turns on a transverse axis between a lateral articulation (A, *a*) and a mesal articulation (*b*), but the mandible has lost the mesal articulation and has acquired a secondary anterior lateral articulation (B, *c*), so that the jaw swings transversely on a longitudinal axis. Each mandible of a generalized insect has four muscles, two arising dorsally on the cranial wall, one on the tentorium, and one at the base of the hypopharynx, but in the higher insects the tentorial and hypopharyngeal muscles have been lost. The two persisting cranial muscles, being attached on opposite sides of the axis of the jaw, then become, one, an *adductor* (*ad*), the other, an *abductor* (*ab*).

The mandibles of the honey bee differ in shape and relative size in the three castes. In the worker the mandible (fig. 23 A, *Md*) is wide at the base, narrowed through the middle, and widened again distally in a flattened expansion with a rounded margin. It has a

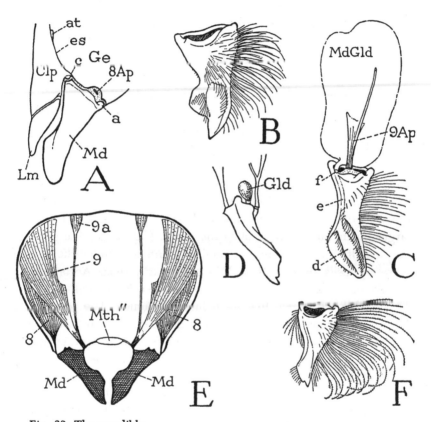

Fig. 23. The mandible.

A, left mandible of a worker, articulated on head. B, right mandible of queen, mesal. C, right mandible of a worker and its gland, mesal. D, right mandible and gland of a drone, anterior. E, diagrammatic cross section of head with mandibles and their muscles, anterior. F, right mandible of drone, mesal.

a, primary cranial articulation of mandible; *c,* clypeal articulation of mandible; *d,* mesal depression of mandible; *e,* groove of mandible; *f,* orifice of mandibular gland. For explanation of abbreviations see page 79.

narrow, flexible membranous attachment on the lower margin of the gena but is articulated anteriorly (*c*) on the lateral angle of the clypeus and posteriorly (*a*) on the gena. The obliquity of the axis between the two articular points causes the jaw to turn posteriorly as well as mesally when it is adducted. The fibers of the abductor muscle (E, 8) spread over the side wall of the head behind the compound eye and converge upon a stalked apodeme arising from the articular membrane at the outer side of the mandible base (A, 8Ap). Most of the fibers of the much larger adductor muscle (E, 9) arise behind the eye above the abductor fibers and on the back of the head and are attached on a thick apodeme, but a small group (9a) arises on the top of the head and is inserted on a long, slender branch of the apodeme. The common stalk of the adductor apodeme arises at the inner side of the base of the mandible.

The inner face of the expanded distal part of the worker mandible (fig. 23 C) is somewhat concave and is traversed obliquely by a channel (d) fringed on both sides with hairs. The channel is continued into a groove (e) that runs upward to an orifice (f) at the inner side of the base of the jaw, which is the outlet of the large *mandibular gland* (*MdGld*). In the worker bee the gland extends upward to the level of the antennal bases (fig. 13 A); it is even larger in the queen (B), but in the drone (fig. 23 D) the gland is very small. The walls of the saclike gland are an epithelium of secretory cells lined with a thin cuticular intima. According to Heselhaus (1922), the cells are penetrated individually by fine canals from the intima that open into the lumen. The function of the manibular gland of the bee is not definitely known (see discussion by Ribbands, 1953), but corresponding glands in other insects probably have some function in connection with the digestion of food. Since the apodeme of the adductor muscle of the mandible in the bee lies close against the mesal surface of the gland (fig. 23 C), it is possible that the adduction of the mandible effects an accompanying discharge of the gland secretion.

The mandibles of the worker bee, aside from being handy tools for any kind of work that requires a pair of grasping organs, are put to various specific uses, including the ingestion of pollen grains for food, the manipulation of wax in comb-building, support of the base of the extended proboscis during feeding on liquids, and maintenance of the flexed proboscis in place when the latter is folded behind the head. Passively the opposed concave inner surfaces of the partly opened mandibles form a conduit for the discharge of nectar and honey or brood food.

The mandible of the queen (fig. 23 B) is about the same length as that of the worker, but it is wider at the base and bilobed distally, the outer lobe being a strong, toothlike projection with a concave inner surface. There is, however, no groove from the orifice of the mandibular gland. The outer surface of the queen mandible is clothed with more numerous and longer hairs than is that of the worker mandible, and in both castes of the female the mandibular hairs are unbranched. The drone mandible (F) is smaller than the mandible of either the worker or the queen. Its distal part is narrow, provided with a small apical tooth, and has a mesal depression from which a faintly marked groove leads up to the base. The hairs of the drone mandible are particularly long and numerous and, in contrast to those of the female mandible, are nearly all of the plumose variety.

THE PROBOSCIS

The proboscis of the bee is a composite organ formed of the maxillae and the labium. In a literal sense, however, it is not an organ, since it becomes a functional unit only when the component parts are brought together in such a way as to form a tube through which liquids may be drawn up to the mouth by the action of the cibarial pump. When not in use, the parts are separated and folded back behind the head (Fig. 24 E). When the parts are artificially spread out (A), the median component as thus seen, ending in the long hairy "tongue" (*Gls*), is the *labium*, the lateral components are the *maxillae*. The three parts are suspended from the membranous floor of the proboscis fossa (*PF*) on the back of the head and are united by a basal V-shaped sclerite known as the *lorum* (*Lr*).

The bee's proboscis gives us an excellent example of how a group of structures that are fundamentally simple in their generalized condition can be integrated and individually elaborated to form a complex organ capable of various activities of which the component parts were originally quite incapable. The insects are particularly endowed for the development of skeletomuscular mechanisms because of the attachment of their muscles on the cuticular layer of the integument, which becomes sclerotized in specific areas to form plates comparable to the movable parts of a machine. Changes in the sclerotization, therefore, can give in different insects entirely different movements produced by the same muscles. More-

59

over, the elasticity of the skeleton in many cases obviates the necessity for antagonistic muscles, the muscular pull in one direction being opposed by skeletal elasticity in the other, so that movement in two directions is accomplished by one muscle instead of two

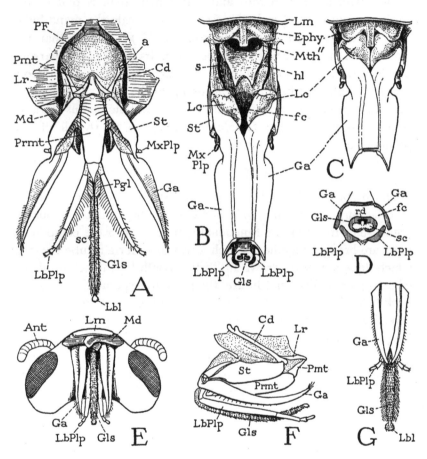

Fig. 24. The proboscis of a worker.

A, labium and maxilla suspended from back of head, with parts artificially separated. B, base of protracted proboscis, anterior, food canal open. C, proboscis with maxillary lobes retracted against epipharynx, closing the food canal. D, cross section through distal part of proboscis. E, proboscis and mandibles folded behind head. F, folded position of parts of labium and maxilla, left side. G, distal parts of proboscis in functional position, tongue protracted, anterior.

a, cranial articulation of cardo; s, suspensory rod of labium. For explanation of abbreviations see page 79.

60

muscles. Animals have seldom developed anything entirely new in their anatomy after the fundamental pattern of their organization was once established in their ancestry, but they have an unlimited potentiality for remodeling and readapting old parts for new purposes. There can be little doubt that the mouth parts of insects have been evolved from three pairs of segmental appendages that were originally designed for locomotion.

The larva of the bee gives us no information concerning the evolution of the maxillae and the labium of the adult. The feeding organs of the larva are in no sense primitive; they are greatly simplified in form and structure, but they are in this way adapted to the needs of the larva. The larval labium has a broad base on the underside of the head (fig. 18 D, *Lb*) united laterally with the simple, slender maxillae (*Mx*). Distally the labium projects in a free lobe (B, *Lb*) with which the hypopharynx (A, B, *Hphy*) is united dorsally. On the end of this labiohypopharyngeal lobe is the slitlike opening of the silk glands between the semitransparent lips of the spinneret (*Spn*). The complex parts of the adult proboscis are formed entirely *within* the larval maxillae and labium (E), but the simple larval organs give us no clue to the identity of the parts of the adult organs. The bee larva is not merely an immature bee; it is a creature with an independent organization, designed to live in a cell and to be fed by adult attendants.

The Maxillae— The maxilla of a generalized insect (fig. 22 C) consists of a long basal part (*Cx*) corresponding with the coxa of a leg (A, *Cx*), a segmented *palpus* (C, *Plp*) representing the telopodite of a leg (A, *Tlpd*), and of two coxal endites, the *galea* (C, *Ga*) and the *lacinia* (*Lc*), arising mesad of the base of the palpus. The coxal base of the appendage is subdivided by an elbow joint into a proximal *cardo* (*Cd*) and a distal *stipes* (*St*). These two parts are not primary segments of the appendage: they have a common opening into the head, no muscles go between them, and the only articulation of the maxilla with the head is that on the base of the cardo (*a*). The elbowlike stipito-cardinal joint merely allows the maxilla to be protracted and retracted on the cardinal articulation. The cranial muscles of the maxilla are attached on the cardo; muscles from the tentorium are attached usually within both the cardo and the stipes. The palpus has a pair of basal muscles, the

61

galea and lacinia have only one muscle each, all arising in the stipes. Keeping these few facts in mind will enable us to identify the parts of the greatly modified and specialized maxilla of the bee.

The maxilla of the adult honey bee retains all the parts and most of the muscles of a generalized maxilla, though the shapes and proportional sizes of the parts are very different. In an extended maxilla (fig. 24 A) we can readily identify the suspensory basal rod (*Cd*), articulating on the margin of the proboscis fossa, as the cardo and the large, boat-shaped section (*St*) depending from the cardo as the stipes. At the end of the stipes arises laterally a very small, two-segmented palpus (*MxPlp*), and directly continued from the stipes is a long thin tapering blade (*Ga*), which is the galea. The galea is traversed throughout its length by a prominent midrib and is narrowed at its base (fig. 25 D), where it becomes continuous with a small triangular plate, or *subgalea* (*Sga*), set on the outer side of the end of the stipes. A representative of the lacinia of a generalized maxilla is not at once apparent on the bee's maxilla, but arising on the end of the stipes mesad of the subgalea is a prominent membranous lobe (A, D, G, *Lc*). Since a muscle (A, F, G, *16*) from the stipes is attached on a small leverlike rod (F, G, *lvr*) in the base of this lobe, there is little doubt that the lobe is the greatly modified lacinia. The lacinial lobe is the *Segelhalter* of Wolff (1875), who termed the muscle the *Spanner des Segelhalters*. The same apparatus variously developed is present also in other Hymenoptera. The lacinial lobes of the bee's maxillae play an important part in closing the temporary food canal of the proboscis, as will later be shown.

Though the bases of the maxillae are freely suspended in the membranous floor of the proboscis fossa, they have a firm attachment on the cranium by means of the cardines. Each cardo articulates by a small condyle near its base (fig. 25 A, *a*) with a process on the margin of the subgenal inflection of the head that forms the corresponding lateral wall of the fossa (fig. 24 A). The extreme end of the cardo projects a little beyond the articular condyle, and on it is inserted the single cardinal muscle (fig. 25 A, *10*) from the head wall. The contraction of this muscle serves to swing the entire maxilla ventrally on the cardinal articulation (*a*). The maxilla is more strongly protracted, however, by three large muscles inserted on its posterior part (*11, 12, 13*), which arise on the extreme anterior end

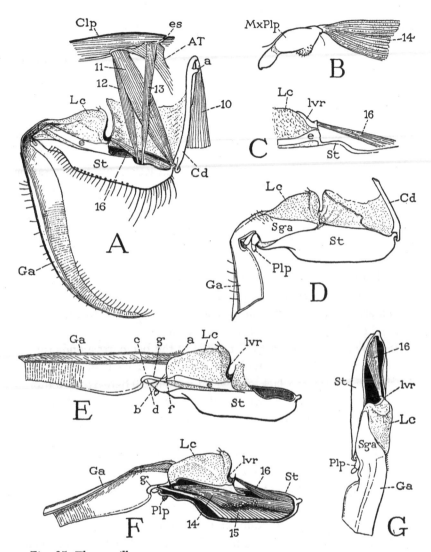

Fig. 25. The maxilla.

A, right maxilla of worker, mesal, with muscles. B, left maxillary palpus and muscles. C, lacinial lobe and lever of right maxilla of a drone, mesal. D, base of left maxilla of a worker, lateral. E, right maxilla with galea extended, mesal. F, same, galea partly deflexed. G, stipes and base of galea of right maxilla, anterior.

a, cranial articulation of cardo; *e,* sclerite at base of lacinia. For explanation of abbreviations see page 79.

of the anterior tentorial arm of the same side (*AT*). The pull of these muscles flattens the angle between the stipes and the cardo, and thus effects a protraction of the stipes on the cardo. This mechanism for the protraction of the maxilla is common to all insects with generalized mouth parts. Because of the close connection of the labium with the maxillae by means of the lorum (fig. 24 A, *Lr*), the protraction of the maxillae protracts also the labium, i.e., the entire proboscis. On the other hand, the maxillae of the bee have no retractor muscles, such as are present in most other insects, and the proboscis as a whole depends for its retraction on the retractor muscles of the labium (fig. 27 A, D, *17*). In this way the maxillae and the labium are interdependent on each other as parts of a single apparatus.

Within the stipes are the muscles of the palpus, the galea, and the lacinia. The single palpus muscle (fig. 25 B, F, *14*) is very large considering the small size of the palpus. The galea has a large muscle (F, *15*) attached on its base. The muscle of the lacinia (*16*) by its attachment on the lacinial lever (*lvr*) serves to elevate the lacinial lobe and maintain it in a taut condition.

It is thus to be seen that the maxilla of the bee corresponds fundamentally in its structure with that of the maxilla of a generalized insect, except for the lack of several muscles. There is no tentorial muscle to the cardo, the small palpus has only one muscle instead of the usual antagonistic pair of muscles, and the usual cranial retractor of the maxilla inserted at the base of the lacinia is absent. The mechanism of the maxilla in the proboscis complex, however, depends on special minor features in the skeletal parts that control and modify the action of the muscles in such a way that entirely new movements are produced, as will later be shown.

The Labium— The labium, or under lip, of the insect is formed in its development by the union of the second pair of embryonic maxillary appendages (fig. 10, *2Mx*), which in most other arthropods remain separate in the adult. The insect labium clearly shows its double origin in its structure. The body of the organ is divided transversely into two parts (fig. 22 D), the distal one of which corresponds with the stipites of a pair of maxillae, since it carries two palpi and four apical lobes. The apical lobes of the labium are termed the *glossae* (*Gls*) and the *paraglossae* (*Pgl*); together they constitute the *ligula* (*Lig*). The basal part of the labium in most insects is a plate on the back of the head, from which the stipital part is freely suspended. The two major parts of the body of the labium have commonly been known as the "mentum" and the "submentum," but in some insects a third part intervenes, and the name "mentum" was first given to it. To avoid the resulting confusion, the names *prementum* and *postmentum* have recently come into use for the parts of a two-part labium (D, *Prmt*, *Pmt*), and the postmentum may then be divided into a mentum and a submentum if the labium has three parts. It must be admitted, however, that the word "mentum" (meaning "chin") is anatomically incongruous in its application to any part of the labium (meaning "lip").

The labium of the honey bee (fig. 24 A) preserves all the parts of a generalized labium. The body of the organ includes an elongate prementum (*Prmt*) and a small triangular postmentum (*Pmt*). Diverging from the end of the prementum are the slender, segmented palpi (*LbPlp*), and between their bases arise the ligular lobes, including the long, median "tongue" (*Gls*), which is evidently the combined glossae, and a pair of short paraglossae (*Pgl*). The postmentum and the base of the prementum are contained in the membranous wall of the proboscis fossa between the bases of the maxillae, but the labium has no direct connection with the cranial walls. The narrowed base of the postmentum is set closely in the angle of the lorum (*Lr*), and the divergent arms of the latter are articulated on the distal ends of the maxillary cardines. The labium and the maxillae are thus yoked together so that the whole proboscis becomes protractile and retractile on the cranial articulations of the cardines.

Each slender labial palpus (fig. 24 A, *LbPlp*) consists of four distinct segments and a small membranous basal lobe by which the palpus is attached to the prementum. The basal lobe may be termed a *palpiger* (fig. 28 A, *Plg*), but it is possible that it is the basal part of the palpus itself. A narrow sclerotic bridge in its anterior wall connects the long proximal segment of the palpus with the prementum, and in the posterior wall is an elastic bar (*n*) on which is attached the single extrinsic muscle of the palpus (*21*). The bar, as will be shown later, has an important function in the flexion of the palpus.

The ligular lobes of the labium, glossa and paraglossae, have a common base on the end of the prementum (fig. 26 B, C), but they are supported anteriorly (B) by a pair of *ligular arms* (*h*) from the margins of the premental plate and posteriorly (C) by a triangular *subligular plate* (*k*) extended from the end of the prementum. The ligular arms (B, *h*) converge past the salivary exit (*slvO*), are then continued as a pair of slender rods (*i*) that support the bases of the paraglossae, and finally curve inward to end in a pair of opposed points articulating with the base of the glossal tongue (*Gls*).

The long, densely hairy tongue of the bee (fig. 24 A, *Gls*) gives no evidence of its having been formed by the union of a pair of glossae, except perhaps in the fact that it is split posteriorly by a deep groove (*sc*), but its median position and its basal musculature leave no doubt of its glossal origin. The tongue has a closely ringed structure in which narrow transverse bands of sclerotization bearing the hairs alternate with bare membranous spaces (fig. 26 D), so that the organ is both flexible and contractile. In cross section the tongue is transversely oval. On the anterior side of its base is a bonnet-shaped sclerite (B, H, *j*) decurved abruptly before the exit orifice of the salivarium (B, *slvO*) and produced distally in a tapering extension on the tongue surface. The tips of the ligular arms of the prementum (B, H, *i*) hold the tongue firmly in place while they allow it to swing freely in a vertical plane on the axis between their opposing points. Distally the tongue tapers somewhat and ends with a small spoon-shaped lobe, termed the *labellum* (fig. 24 A, *Lbl*), which has a smooth convex posterior surface but is armed on its free margin and on its concave anterior surface with small branched hairs (fig. 26 G). The labellum of the bee's tongue is not comparable with the labellar lobes of a fly's proboscis, since the latter represent the labial palpi.

The narrow median groove that traverses the posterior side of the tongue expands within the tongue into a wide channel (fig. 24 D, *sc*). The lips of the channel are closed by dense fringes of small hairs directed medially and distally (fig. 26 G), and the channel opens on the narrowed base of the smooth posterior surface of the labellum. Running through the inner wall of the glossal channel is a slender, strongly sclerotic but elastic rod (fig. 24 D, *rd*), which is itself grooved on its inner surface. The groove of the rod also is closed by fringes of small hairs. At the base of the tongue the lips of the glossal channel spread apart (fig. 26 C) and

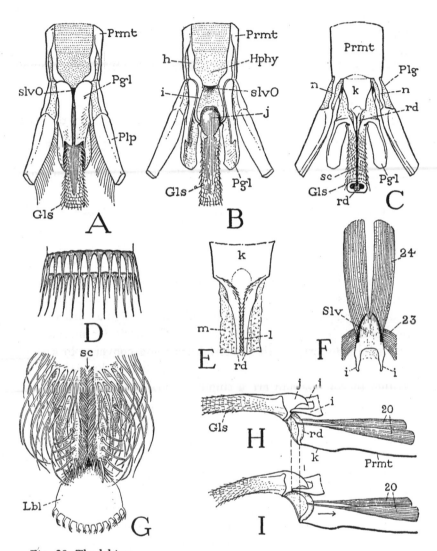

Fig. 26. The labium.

A, end of prementum with bases of terminal lobes, paraglossae in natural position, anterior. B, same, paraglossae spread out. C, same, posterior. D, two bands of glossal hairs, anterior. E, base of glossal rod, posterior. F, posterior wall of salivarium and muscles. G, end of glossal tongue, posterior. H, diagram of protracted tongue. I, tongue partly retracted.

k, distal plate of prementum. Other alphabetical lettering is explained in text. For explanation of abbreviations see page 79.

expose the rod (*rd*), which is here seen to be directly continued from the apex of the triangular subligular plate (C, E, *k*) on the end of the prementum.

The paraglossae are thin, elongate lobes arising from the common ligular base on the prementum at the sides of the tongue (fig. 26 A, B, C, *Pgl*), where they are supported on the ligular arms of the prementum (B, *h*). Their concave inner surfaces clasp the base of the tongue (A) and underlap its posterior surface, closing over the proximal end of the tongue channel. Between the bases of the paraglossae on the anterior surface of the labium is the opening of the salivarium (A, B, *slvO*).

The musculature of the bee's labium consists of extrinsic muscles from the head, muscles confined to the prementum, and muscles within the basal segments of the palpi. The head muscles of the labium, as in most other insects, include two pairs of muscles, one pair inserted distally on the prementum, the other proximally. The anterior muscles (fig. 27 A, D, *17*) arise on the dorsal wall of the cranium and insert on the distal ends of the ligular arms of the prementum (*h*, shown infolded at A). The posterior muscles (D, *18*) arise anteriorly on the tentorial arms and converge to a single tendon attached on the proximal extremity of the prementum. Within the prementum are a single muscle of each palpus, a pair of glossal muscles (A, *20*) attached on the basal part of the tongue rod, a pair of muscles (*19*) attached on the ligular arms (*h*), which evidently are the paraglossal muscles of a generalized labium, and, finally, muscles of the salivarium (*23, 24*). The function of the muscles and the operation of the labial mechanisms will be described in the next two sections of this chapter.

It is of interest to note that the labium of the bee larva (fig. 18 D, *Lb*) has nothing of the structure of the adult labium. It is a simple lobe without palpi or ligular lobes and is united with the hypopharynx. Yet within the larval labium (E) are developed all the parts of the pupal labium, which are those of the adult. It is clear that the larval labium is specialized by simplification for the specific purposes of the larva; it does not serve as a mold for the adult labium.

The Labial Glands— In most insects a pair of glands lies in the ventral part of the thorax, the ducts of which enter the head and unite in a common exit duct that opens generally in the pocket

68

between the base of the hypopharynx and the prementum of the labium, though in piercing insects the duct may traverse the hypopharynx and open at the tip of this organ. These glands are commonly known as the "salivary glands" of the adult insect, be-

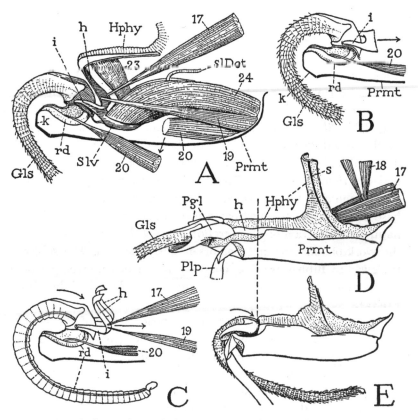

Fig. 27. Labial muscles and the retractor mechanism of the ligula.

A, lengthwise section of prementum of worker; ligula and salivarium retracted by pull of muscles 17 and 19 on ends of ligular arms (h). B, detail of base of retracted and recurved tongue. C, diagram of retraction and curvature of tongue. D, prementum, with ligula protracted. E, same, ligula retracted, tongue fully recurved.

Alphabetical lettering is explained in text. For explanation of abbreviations see page 79.

cause, with the usual opening of the duct at the base of the labium, their secretion is discharged into the preoral cavity where it comes into contact with the food. In blood-feeding insects, however, the secretion has strong irritating properties and may be an anticoagulant of the blood, and in lepidopterous and hymenopterous larvae the glands secrete liquid silk. It is impossible, therefore, to give a consistent functional name to these glands, but anatomically they are *labial glands,* because they are developed as ectodermal ingrowths on the labial segment of the head.

Glands similar to those in the thorax of other insects are present in the honey bee. They lie ventrally in the anterior part of the thorax (fig. 70, *ThGld*), their ducts run forward into the head and unite, but the common duct here receives two lateral ducts from another pair of glands in the back of the head (*HGld*). The single terminal duct finally opens into the salivary pocket, or salivarium (fig. 17 B, *Slv*), at the base of the labium. The two sets of glands in the labial gland system of the bee (fig. 21 C) are usually distinguished as the *head glands* (*HGld*), or *postcerebral glands,* and the *thoracic glands* (*ThGld*). For convenience they may be termed the *salivary glands,* since their secretion discharged on the base of the proboscis must come in contact with the food, but the nature and function of the secretion are not definitely known, since the work of different investigators is not in accord. The subject is reviewed by Ribbands (1953, chap. 7); the glands are well portrayed by Bugnion (1928); their finer structure is described by Kratky (1931).

The head glands lie against the posterior walls of the cranium (fig. 20, *HGld*). In the worker each gland (fig. 21 C) consists of a loosely arranged mass of small pear-shaped bodies (D) with individual ducts that unite irregularly with each other and eventually come together in a single duct that joins the common median duct from the thoracic glands. The head glands are equally developed in the worker and the queen but are vestigial or absent in the drone.

Each thoracic gland (fig. 21 C, *ThGld*) consists of a mass of many-branched glandular tubules opening into short collecting ducts (E) that unite in several major ducts which end in a saclike reservoir (C, *Res*) at the anterior end of the gland. The final outlet duct proceeds forward from the reservoir, and, as already noted, is joined by the ducts from the glands in the back of the head. The thoracic glands are present in the worker, queen, and drone and represent the usual "salivary glands" of other adult insects.

Since the common duct of the labial glands opens through the salivarium onto the base of the labium, the secretion must be conveyed within the paraglossae to the back of the tongue, where it enters the tongue canal to be finally discharged upon the labellum and mixed with the food being taken up through the channel of the proboscis. Although investigators differ as to the presence or the number of enzymes in the secretion of the labial glands, it would seem that an important ingredient of the bee's saliva should be the invertase that converts the cane sugar of nectar into the levulose and dextrose of honey. The change takes place largely in the crop, which has no glands, but according to Park (1954), "the inversion process begins while the nectar is being gathered and carried to the hive." Inglesent (1940) says the thoracic glands of bees of all ages contain much invertase, though previous writers had reported little or none present.

In the bee larva the head glands are not yet developed. The thoracic glands are represented by a pair of long, slender, convoluted tubes (fig. 68 A, *skGld*) extending back to the sixth abdominal segment, which are the larval *silk glands*, but the thoracic glands of the adult are developed from them. In the bee embryo, according to Nelson (1915), the silk glands are formed as a pair of tubular ingrowths of the epidermis behind the bases of the rudiments of the second maxillae. As the tubules lengthen, the second maxillae unite to form the labium, and at the same time the orifices of the tubules come together in the base of the labium to form a common exit duct, which, Nelson says, "ultimately opens at the tip of the labium." The larval structure that Nelson calls the "labium," however, as already noted, is the combined labium and hypopharynx (fig. 18 B). The exit canal of the larval silk glands, therefore, is the salivarium between the hypopharynx and the labium, into which opens the common salivary duct of the adult. In a caterpillar the salivarium forms the "silk press."

The silk glands of the bee larva, as described by Anglas (1901), attain their highest development just at the time the larva is ready to spin its cocoon. This finished, the glands rapidly degenerate during the first part of the pupal period: the cells lose their outlines and become irregular masses of vacuolated protoplasm, in which the nuclei break up into fragments that quickly degenerate. In a pupa 70 hours after sealing, Oertel (1930) says no trace of the

71

silk glands remains. The thoracic glands of the adult, according to Schiemenz (1883), are formed inside the basement membrane of the first part of the larval silk gland, which probably means that the adult glands are developed by renewed growth from remnants of the larval glands. The head glands are secondary outgrowths from the duct of the thoracic glands.

The Salivarium— In the adult bee the salivarium functions as a salivary ejection pump, or "salivary syringe." It is derived from the pocket of the preoral cavity of a generalized insect (fig. 17 A, *Slv*) between the hypopharynx and the labium, into which the salivary duct opens. By a union of the lateral edges of the posterior wall of the hypopharynx with the anterior wall of the prementum, the salivarium of the bee (B, *Slv*) becomes a closed chamber opening on the distal end of the prementum between the bases of the paraglossae (fig. 26 A, B, *slvO*). The posterior, labial wall of the chamber is a rigid, slightly concave plate (figs. 26 F, 27 A, *Slv*). The anterior, hypopharyngeal wall is flexible and elastic and gives insertion to a pair of dilator muscles (23) convergent from the outer wall of the hypopharynx. Attached on the lateral margins of the salivarium are two huge muscles (24) arising in the base of the prementum. These muscles evidently act as compressors of the salivarium by flattening the chamber after its expansion by the dilator muscles. The salivarium thus probably sucks the gland secretion from the duct and forcibly expels it from the distal opening into the space between the paraglossae at the base of the tongue (fig. 26 A, B). The muscles of the salivarium, however, can be effective only in the protracted state of the proboscis; when the tongue is retracted, as shown at A of figure 27, the large compressor muscles (24) are slack and clearly nonfunctional.

The Proboscis as a Feeding Organ— In the functional feeding position of the proboscis the maxillary galeae and the labial palpi are brought together around the tongue to form a tube through which liquids may be drawn up to the mouth by the sucking action of the cibarium. The base of the proboscis is now held between the extended mandibles, the concave inner surfaces of which fit snugly over the midribs of the galeae. The galeae (fig. 24 B, D, *Ga*) overlap anteriorly to form the roof of the food canal (D, *fc*); the palpi (*LbPlp*) and their marginal hairs close it posteriorly. The basal half of the tongue (*Gls*) is thus enclosed in the food canal, but a variable length of its distal part projects freely beyond it (G). The tapering ends of the galeae converge in front of the tongue; the small apical segments of the labial palpi diverge laterally and probably serve as sensory outposts.

When the proboscis is extended (fig. 24 B), the upper end of the food canal (*fc*) is wide-open, exposing the functional mouth (*Mth″*) and the biblike hypopharyngeal lobe (*hl*) depending from the lower margin of the suboral plate. During feeding, however, the base of the proboscis is drawn up to the head (C); the lacinial lobes of the maxillae (*Lc*) are now pressed close against the epipharynx (*Ephy*) with the median keel of the latter between them. The food canal thus becomes a completely closed channel leading from the distal opening between the tips of the galeae and palpi up to the mouth.

A worker bee feeding on honey plunges the end of the proboscis into the liquid; the free part of the tongue is bent back on the surface of the honey and at once begins a back and forth movement. The honey drawn into the proboscis by the action of the tongue rapidly ascends the food canal, presumably by the suction of the cibarial pump, and is drawn into the mouth over the lobe of the hypopharynx, which forms a ramp from the floor of the food canal of the proboscis to the wide-open oral aperture leading into the cibarial chamber.

The method by which bees feed on sugar is described as follows by Jucker-Piédallu (1934). The end of the tongue is recurved, so that the anterior spiny surface of the labellum can be applied against the sugar, and probably it acts as a rasp. At the same time a clear liquid, apparently saliva, is discharged from the canal of the tongue onto the labellum and by means of the fringe of hairs on the latter is spread over the sugar. By friction and by the solvent action of the saliva, the sugar is gradually transformed into a syrup, which the bee readily takes up.

The movements of the tongue made by a feeding bee are so varied and so rapid that the tongue appears to be itself endowed with mobility. The tongue as a whole lengthens and contracts, and the exposed part moves about in all directions. The tongue movements, however, are all produced by the single pair of muscles attached on the base of the glossal rod. The proximal end of the rod curves backward from the base of the tongue (fig. 26 H, *rd*) to its attach-

73

ment on the subligular plate (*k*) of the prementum. The rod muscles (*20*) are attached by tendons, not directly on the rod, but immediately at the sides of its recurved basal part. Since the base of the tongue is braced against the prementum by the ligular arms (*i*) of the latter, the contraction of the rod muscles increases the basal curvature of the rod (I) and pulls the rod back from the tongue as far as the membranous walls of the tongue canal will permit. Owing to the compressibility of its ringed wall the tongue itself is thus shortened, and the shortening is greatest at its distal end. On relaxation of the muscle tension the tongue again extends, evidently by elasticity of the rod and probably of the tongue walls themselves. The attachment of the muscles at the sides of the base of the rod enables the muscles, acting singly or in opposition, to twist the rod on its long axis, and by this action on the base of the rod the various lateral movements of the distal part of the tongue are produced.

The canal of the tongue probably plays no part in the ingestion of food liquids. The canal is completely shut in by dense fringes of marginal hairs. It opens distally on the base of the posterior surface of the labellum (fig. 26 G) and proximally at the base of the tongue (C) between the paraglossae, and it has no direct connection with the sucking apparatus. On the other hand, the closure of the paraglossae behind the tongue puts the tongue channel in communication with the opening of the salivarium between the bases of the paraglossae on the anterior surface of the labium (A, *slvO*). The saliva forcibly expelled encounters at once the steeply declivous surface of the basal plate of the tongue (B, *j*), by which it must be deflected in two divergent streams into the concavities of the mesal surfaces of the paraglossae and within the latter be conducted around to the posterior surface of the tongue. Here its only outlet would be into the open proximal end of the tongue canal (C). Through the canal the saliva should then be conveyed to the tip of the tongue and discharged upon the smooth undersurface of the labellum, where it would mix with the food during feeding. So far as known to the writer no direct observation has been made on the course of the saliva in a living bee, but the anatomical relations of the parts here described would seem to leave no other exit available than the glossal canal. If the minute groove of the tongue rod itself has a specific function, it is not

74

known. Since the marginal hairs that close over both the canal of the tongue and the groove of the tongue rod are directed distally, it would appear that the tongue does not in any case serve for the intake of liquids. Particularly well guarded is the distal opening of the tongue canal at the base of the labellum (fig. 26 G).

The passageways around the base of the tongue within the paraglossae, in addition to serving as salivary conduits, are probably used also in the "ripening" of nectar into honey. Park (1925) has described and illustrated the discharge of a drop of nectar from the mouth of a bee in the hive and its accumulation in the angle between the posterior surface of the prementum and the reflexed distal parts of the proboscis. The nectar globule is several times sucked back into the mouth and redischarged. Though Park does not explain how the nectar gets behind the tongue and is again returned to the mouth in front of the tongue base, its only course must be that between the tongue and the paraglossae. Probably also the nectar flows down the tongue channel itself, the walls of which are eversible. Park suggests that some enzyme is probably added to the nectar from glands in the head; if so, the glands discharging from the salivarium would be the most likely source. The "ripening" process consists of the evaporation of water from the nectar. After it is finished, the bee deposits the honey in a cell of the comb from between her opened mandibles.

The Folding of the Proboscis— When the proboscis is not functionally active, its distal parts, including the maxillary galeae, the labial palpi, and the tongue, are turned back against the posterior side of the head (fig. 24 E), where they are held in place by the mandibles clasped beneath them, while the labrum clamps down on the mandibles. The folding of the proboscis as it is done by the bee appears to be a very simple matter, but the folding mechanisms depend on some complex anatomical adjustments in the parts involved.

A simple backward flexion of the maxillary galea is impossible because the larger part of the galeal blade extends in a vertical plane from the end of the stipes (fig. 25 E, *Ga*). Where the galea joins the supporting subgaleal plate (D), it is abruptly narrowed. A rodlike thickening (E, *g*) of the posterior margin of the narrow part articulates proximally on the laterally curved end (*f*) of a long, slender sclerite (*e*) along the mesal wall of the stipes at the side of the lacinial lobe. The flexor muscle of the galea (F, *15*) is attached by a tendon on the marginal rod (*g*). The pull of the muscle depresses the galea and at the same time turns it laterally because of the obliquity of the line of bending (E, *a-b*) between the galea and the subgalea. A second line of bending (*c-d*) beyond the first and oblique in the opposite direction, however, soon counteracts the lateral movement, and the continuing pull of the muscle now turns the galea straight back and finally folds it up against the stipes. Extension of the galea evidently results automatically on relaxation of the muscle by elasticity of its basal part, which is continuous with the firmly affixed subgalea.

The folding of the labial parts involves separate mechanisms for the tongue and the palpi. Prior to the flexion of the tongue the bases of the tongue and the paraglossae (fig. 27 D) are retracted into the end of the prementum (E). The retraction is produced by the infolding of the ligular arms (A, C, *h*) of the prementum by the pull of the muscles (*17, 19*) attached on the distal ends of the latter. By the retraction of the tongue the tongue rod (*rd*) is drawn back in a wide loop from its attachment on the end of the sub-ligular plate (A, B, *k*). Since the rod traverses the posterior side of the tongue (C) in the ample membrane of the tongue canal, the contractile and flexible tongue is not only shortened but sharply curved in a posterior direction. Along with the retraction of the tongue and paraglossae, the salivarium is also pulled into the prementum (A, *Slv*), and its muscles (*23, 24*) become slack and temporarily functionless. Inasmuch as there are no muscles for the protraction of the indrawn parts, their restitution to the functional position must depend on elasticity of the ligular arms and the sub-ligular plate when the muscles relax, unless some blood pressure is exerted through the labium.

The flexing mechanism of the labial palpus is somewhat similar to that of the maxillary galea. The palpus, as already noted, is attached to the prementum by a short basal part, which for convenience we have called the palpiger (figs. 26 C, 28 A, *Plg*). The walls of the palpiger are membranous except for a narrow sclerotic band in the anterior wall and a slender flexible and elastic rod (fig. 28 A, D, *n*) in the posterior wall, which is continuous from the prementum into the midrib of the palpus. The single basal muscle of the palpus (A, *21*) is attached by a long tendon to this

76

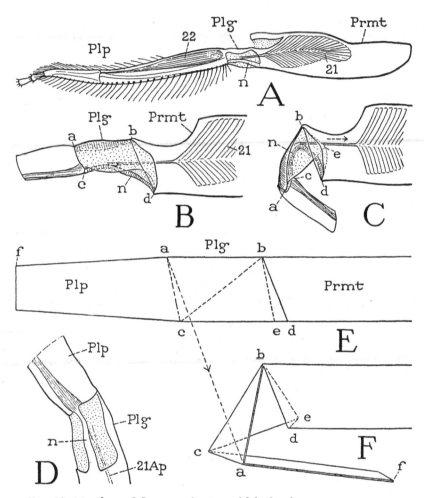

Fig. 28. Muscles and flexor mechanism of labial palpus.

A, right palpus and prementum, with palpal muscles, mesal. B, base of palpus, in position of extension. C, same, palpus flexed. D, same, posterior, showing elastic rod (*n*) in palpiger, and attachment of flexor tendon (*21Ap*). E, F, diagrams of flexing of palpus on prementum, illustrated with piece of stiff paper cut and creased as at E and folded as at F.

Alphabetical lettering is explained in text. For explanation of abbreviations see page 79.

posterior rod a little beyond the middle of the latter. The pull of the muscle in contraction curves the flexible rod into the membranous wall of the palpiger (B, C), which action abruptly deflexes the shaft of the palpus (C). The exact movement of the shaft, however, depends on specific lines of bending at the two ends of the palpiger.

The flexing action of the palpus can best be illustrated with a piece of stiff paper cut into the shape shown at E of figure 28 and creased along the lines *a-c, b-c, b-d,* and *b-e.* If the paper is folded outward along the diagonal *b-c,* the triangle *abc* is turned over and directed downward (F). The distal part representing the palpus can now be bent backward on the line *a-c,* but its alignment with the prementum necessitates a compensating fold inward along the lines *b-d* and *b-e.* The distance from *c* to *d* is thus shortened, and the palpal shaft can now be folded parallel to the prementum. In the flexing of the palpus of the bee, the infolding of the palpiger results from the inward curvature of the rod in its posterior wall (C, *n*) effected by the pull of the muscle on the convex side of the rod. Because of the several preformed lines of folding, as illustrated in the paper model, the tension of the single palpus muscle automatically turns the palpus over and bends it backward against the stipes. Since the palpus has no extensor muscle, it evidently resumes the extended position by the elasticity of the palpiger rod on relaxation of the muscle.

In the folded position of the proboscis the open space on its base before the mouth serves as a food trough for the feeding of other bees and as a receptacle for regurgitated nectar delivered by a foraging bee to a receiving bee in the hive. Park (1925) says that a field bee on approaching a receiving bee in the hive "opens her mandibles wide apart and forces a drop of nectar out over the upper surface of the proximal portion of her tongue, the distal portion being folded back under the head." The receiving bee then "stretches out her tongue to full length and sips the proffered nectar." Furthermore, since the hypopharyngeal food glands open on the distal corners of the suboral pate, the secretion must flow down the suboral lobe of the hypopharynx and accumulate at the base of the tongue, where it becomes accessible to other adult bees. When fed to the larvae, however, the royal jelly, or bee milk, is discharged from between the opened mandibles.

78

Explanation of Abbreviations on Figures 17–28

ab, abductor muscle (8) of mandible.

ad, adductor muscle (9, 9a) of mandible.

Ant, antenna.

8Ap, abductor apodeme of mandible.

9Ap, adductor apodeme of mandible.

at, anterior tentorial pit.

AT, anterior tentorial arm.

Br, brain.

Ca, corpus allatum.

Cb, cibarium.

Cc, corpus cardiacum.

Cd, cardo.

Clp, clypeus.

Cx, coxa.

dlcb, dilator muscles of cibarium.

dlphy, dilator muscles of pharynx.

E, compound eye.

Ephy, epipharynx.

es, epistomal sulcus.

fc, food canal of proboscis.

FGld, hypopharyngeal food gland.

For, occipital foramen.

Fr, frons.

frGng, frontal ganglion.

Ga, galea.

Ge, gena.

Gld, gland.

Gls, glossa (tongue).

HGld, head salivary gland.

hl, hypopharyngeal lobe.

Hphy, hypopharynx.

hpl, hypopharyngeal plate.

HS, hypopharyngeal suspensorium.

ibs, infrabuccal sac.

Lb, labium.

Lbl, labellum.

LbPlp, labial palpus.

Lc, lacinia.

Lig, ligula.

Lm, labrum.

Lr, lorum.

lvr, lacinial lever.

Md, mandible.

MdC, cavity where mandible detached.

MdGld, mandibular gland.

Mth', true mouth, opening into stomodaeum.

Mth'', functional mouth.

Mx, maxilla.

MxPlp, maxillary palpus.

O, ocellus.

Oc, occiput.

Oe, oesophagus.

PF, fossa of proboscis.

Pgl, paraglossa.

Phy, pharynx.

Plg, palpiger.

Plp, palpus.

Pmt, postmentum.

Prmt, prementum.

rd, glossal rod.

Res, reservoir of thoracic salivary gland.

RNv, recurrent nerve.

sc, salivary canal of tongue.

Sga, subgalea.

slDct, common salivary duct.

slO, orifice of salivary duct.

Slv, salivarium (salivary syringe, *Syr*).

SO, sense organs.

Spn, spinneret.

St, stipes.

TB, tentorial bridge.

ThGld, thoracic salivary gland.

Tlpd, telopodite.

Tnt, tentorium.

✻ V ✻

THE THORAX

THE thorax is the locomotor center of the insect since it carries the legs and the wings and contains the muscles that move these organs. In most insects it is composed of the first three body segments following the head, which are named the *prothorax,* the *mesothorax,* and the *metathorax.* Each segment bears a pair of legs, but the wings pertain only to the mesothorax and the metathorax. The wings are secondarily acquired organs that the primitive insects did not possess, and as a consequence the thoracic segments of winged insects have been specially modified structurally in adaptation to the flight function imposed upon them. Inasmuch as some insects fly with both pairs of wings, some principally or entirely with the front wings, and some with the hind wings only, the structure of the wing-bearing segments is modified accordingly and may be very different in different groups of insects. In the bees and the wasps the thorax has undergone such an extreme degree of modification that its parts in these insects are not readily identifiable with those of the typical thorax of a more generalized insect. An understanding of the bee's thorax, therefore, will be facilitated by first reviewing the fundamental thoracic structure as it has been deduced from a comparative study of other less specialized insects.

GENERAL STRUCTURE OF THE INSECT THORAX

A thoracic segment may first be analyzed into four topographical surfaces, a dorsal region on the back, a ventral region below between the legs, and a pleural region on each side. The sclerotization of the back in general anatomical nomenclature is the *tergum,* but that of the insect thorax is more appropriately called the *notum*

80

in order that the name may properly combine with the Greek prefixes *pro, meso,* and *meta,* by which the thoracic segments and their parts are distinguished. The ventral sclerotization is the *sternum;* the lateral sclerotizations are the *pleura,* which in a wing-bearing segment lie between the wings above and the legs below. These major skeletal areas, however, are commonly subdivided into smaller parts, which are *tergites, pleurites,* and *sternites.* The legs are primitive segmental appendages of all arthropods; the wings are secondary extensions of the lateral margins of the mesonotum and the metanotum.

The Wing-bearing Notum— The back plate of a typical winged segment (fig. 29, *N*) is marked in its posterior part by a V-shaped groove (*ns*), the apex of which is forward, with the arms diverging

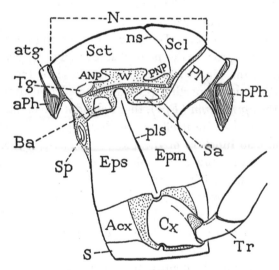

Fig. 29. Diagram of the typical structure of a wing-bearing thoracic segment.
For explanation of abbreviations see pages 92–93.

posteriorly toward the rear angles of the notum. The part of the notum before this *notal sulcus* is called the *scutum (Sct),* that behind it the *scutellum (Scl).* The function of the groove, however, is not to "divide" the notum but to form a strengthening ridge on its inner surface. The ridge evidently serves to increase the elasticity of the notum after it has been buckled down at the ends by the dorsal muscles of flight, the principal movements of the wings in most insects being produced by the alternate arching and flattening of the supporting notal plate. At the anterior margin of the notum is a deep inflection (*aPh*) forming a *phragma* for the attachment of the dorsal muscles of the segment. The phragma is merely an enlargement of the tergal antecosta and usually bears a narrow acrotergite (*atg*) on its anterior lip. On each lateral margin of the notum are an *anterior* and a *posterior notal wing process* (*ANP, PNP*) that support the wing base (*W*). Behind the scutellum of the principal wing-bearing segment of the thorax there may be a distinct plate, the *postscutellum*, or *postnotum* (*PN*), which is the enlarged acrotergite of the following segment, cut off with its phragma from its own segment and transferred to the preceding segment. A segment may thus secondarily come to have an *anterior phragma* (*aPh*) and a *posterior phragma* (*pPh*) by borrowing one from the segment behind.

The Thoracic Pleuron— The pleuron of a wing-bearing segment is cut by a deep groove, the *pleural sulcus* (fig. 29, *pls*), extending upward from the leg articulation to the base of the wing, where it ends in a wing-supporting process. The pleural area before the groove is the *episternum* (*Eps*), that behind it the *epimeron* (*Epm*), each of which may be itself subdivided. Again, as with the notal sulcus, the true function of the pleural sulcus is to form a strong internal ridge that braces the pleuron in its support of the leg at one end and the wing at the other. Above the upper edge of the episternum in front of the wing process is a small sclerite known as the *basalare* (*Ba*), and behind the wing process above the epimeron is a corresponding sclerite termed the *subalare* (*Sa*). These *epipleural sclerites* and the muscles attached on them have important functions in controlling the wing movements. The episternum may be extended downward before the leg base to the sternum, or in some insects there is a special antecoxal plate (*Acx*) between the episternum and the sternum, but again the two parts may be entirely continuous, in which case it is difficult to say where one ends and the other begins.

The Thoracic Sternum— The ventral sclerotization of a thoracic segment is usually more variable in its extent and pattern than is that of the dorsum. Generally it is differentiated into an anterior plate termed the *basisternum*, or *eusternum*, and a posterior plate called the *furcasternum,* or *sternellum.* The term "furcasternum" is derived from the fact that in the higher insects the plate so named bears a Y-shaped endosternal structure, or *endosternum,* known as the *furca.* In the more generalized insects, however, there is merely a pair of sternal apodemes unsupported by a median stalk, so that the term "furcasternum" is not always appropriate. In the bee the sterna of the wing-bearing segments are undivided.

The coxa of a leg (fig. 29, *Cx*) is primarily articulated between the pleuron and the sternum, but in many insects the sternal articulation is lost, and a small sclerite, the *trochantin,* attached dorsally on the episternum, gives a secondary anterior articulation to the coxa.

THE THORAX OF THE HONEY BEE

The thorax of bees and of other Hymenoptera in the suborder Clistogastra, or Apocrita, contains the usual three segments, but it includes also a fourth segment, which is the first abdominal segment of other insects. This added segment is known as the *propodeum.* The deep constriction of the body that separates the abdomen from the thorax is therefore between the primary first and second abdominal segments.

In the thorax of the adult honey bee (fig. 30) the component segments are so thoroughly consolidated that their limits are not evident at first sight. The back is strongly arched from the neck in front to the abdominal petiole behind. The first legs are attached anteriorly, and a long space intervenes between them and the other two pairs of legs, which are attached posteriorly. The wings arise near the center of the lateral aspects of the thorax. Numerous grooves cut the thoracic surface into irregular areas, but three of these grooves are of special significance because they are the dividing lines between the segments. The first intersegmental groove (*1is*) separates the pronotum (*N_1*) from the mesonotum (*N_2*). The second intersegmental groove (*2is*) separates the metathorax from the mesothorax. It runs obliquely upward from the base of the second leg to a point between the wing bases and is then continued upward and posteriorly over the back and down to the leg base on the opposite side. The third intersegmented groove

(*3is*), curving upward and forward on each side from the base of the third leg, separates the propodeal tergum (*IT*) from the metathorax.

It is difficult to see a reason for all the specialization of the thorax of the bee, but in any case reasons are harder to see than

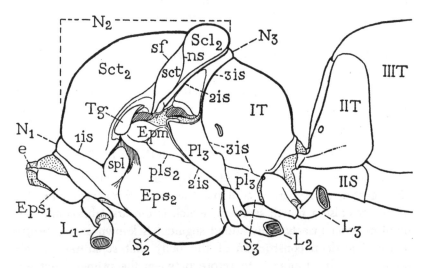

Fig. 30. Thorax and base of abdomen of a worker bee.

e, occipital process of prothoracic episternum. For explanation of abbreviations see pages 92–93.

visible facts. We assume, however, that there is a reason for everything in nature, including the bee's thorax, and there can be little doubt that structural adaptations to flight account for the principal features of the thorax in most winged insects. The condensation of the thoracic segments into a form approaching that of a sphere, as Duncan (1939) says of the thorax of the wasp, clearly gives maximal solidity to the thorax as a whole that enables it to withstand the stress put upon it, particularly by the huge wing muscles of the mesothorax.

The Prothorax— The first legs (fig. 30, L_1) are suspended from the posterior ends of elongate triangular plates that appear to lie in the sides of the neck, but the articulation of the legs on these plates shows that the plates are the prothoracic pleura. When either pleuron is fully exposed (fig. 31 C), there is seen a groove (*pls*) close to its posterior margin extending upward from the coxal articulation, and from its upper end arises a large internal apodemal arm (C, E, *PlA*). This groove, therefore, is the pleural sulcus and divides the propleuron into a large anterior episternum (*Eps*) and a narrow posterior epimeron (*Epm*). Anteriorly the two episterna taper to a pair of peglike processes (*e*) that articulate with the occipital condyles on the back of the head (B). Close to the upper margin of each episternum is a groove (C, *b*) from which is inflected a narrow shelflike apodemal plate (E, *c*) that ends anteriorly in an expansion (*d*) for the attachment of muscles (fig. 15 A, B, *46, 47, 51*).

Ventrally, between the lower margins of the prothoracic episterna is a triangular sternal plate (fig. 31 D, *Bs*) which is the basisternum of the prothorax. Posteriorly the basisternum is connected by a narrow neck with a much smaller furcasternum (*Fs*), which supports internally an elaborate endosternum (B, E, *Endst*) that occupies much of the interior of the prothorax. The endosternum consists of a median support on the furcasternum, a pair of wide divergent arms, and a broad horizontal bridge between the arms, from which arises a pair of long muscle apodemes (E, *51Ap*). Beneath the bridge is a short channel through the endosternum that gives passage to the ventral nerve cords.

The notum of the prothorax (fig. 30, N_1) is entirely separated from the pleurosternal parts of its segment. It is a collarlike plate fitted so closely on the anterior end of the mesothorax that it appears to belong to this segment; its tapering lower ends meet on the venter behind the prothoracic sternum, though they do not unite with each other (fig. 31 A). The pronotum is united with the mesothorax for its own security, since it gives attachment to prothoracic and head muscles and has no part in the mechanism of the mesothorax. On each side the pronotum is produced into a lobe (fig. 30, *spl*) that covers a depression of the mesopleuron, within which is the first thoracic spiracle (fig. 88 B, *1Sp*). In front of the pronotum between the upper edges of the pleural plates is the dorsal wall of the membranous neck.

The Mesothorax— The mesothoracic segment accounts for the major part of the bee's thorax. It is limited in front by the pronotal collar (fig. 30, N_1) and posteriorly by the second intersegmental sulcus (*2is*). The notum (N_2) covers the strongly convex anterior and upper part of the thorax. A transverse groove (*ns*) in its posterior part that runs out laterally into the posterior margin of the notum (fig. 32 A, *ns*) forms internally a strong ridge (B, *NR*), by which the groove is to be identified as the notal sulcus of a more typical thoracic notum (fig. 29, *ns*). The large part of the notum

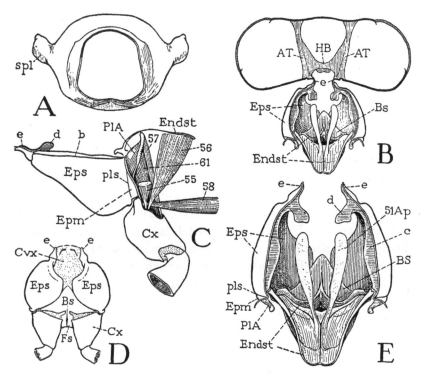

Fig. 31. The prothorax of a drone.

A, pronotum, anterior. B, pleurosternal region of prothorax and horizontal section of head supported on occipital processes of episterna. C, propleuron and base of first leg, left side. D, prothorax and bases of first legs, ventral. E, pleurosternal and endosternal skeleton, dorsal.

b, submarginal groove of episternum; *c*, submarginal apodemal ridge of episternum; *d*, cervical apodeme of episternum; *e*, occipital process of episternum. For explanation of abbreviations see pages 92–93.

before this groove in the bee, therefore, is the scutum (figs. 30, 32 A, *Sct*), and the small posterior part is the scutellum (*Scl*). The mesonotum of the bee, however, is completely divided by a transverse fissure (*sf*) through the posterior part of the scutum, which cuts off a small posterolateral area of the scutum (*sct*) on each side. This division of the mesonotum by a *transscutal fissure* is an unusual feature, but it is essential for the wing movements of the bee, as will be explained in connection with the wing mechanism. The anterior margin of the scutum is deflected to form a small prephragma of the mesothorax (fig. 32, B, *1Ph*). On the posterior parts of the lateral margins of the mesonotum are attached the forewings, each of which is supported on a pair of notal wing processes of the scutum (A, *ANP, PNP*).

The scutellum of the mesonotum includes the prominent swelling on the uppermost part of the back (fig. 30, *Scl₂*), and tapers downward on the sides to the bases of the forewings. The posterior

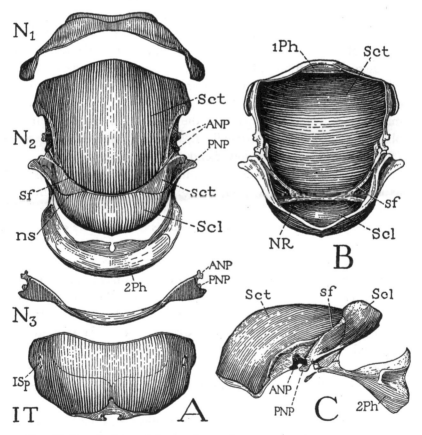

Fig. 32. Dorsal plates of thorax of a worker.

A, pronotum (*N₁*), mesonotum (*N₂*) with second phragma (*2Ph*), metanotum (*N₃*), and propodeal tergum (*IT*), dorsal. B, mesonotum, ventral. C, mesonotum and second phragma, left side.

For explanation of abbreviations see pages 92–93.

scutellar margin supports directly a large postphragma (fig. 32 C, *2Ph*), there being no postnotal plate such as that which ordinarily bears the second thoracic phragma (fig. 29, *PN*) when this phragma is carried by the mesonotum. Moreover, the phragma does not have the usual form of a vertical plate; it is a U-shaped structure (fig. 33 D, *2Ph*) projecting far back through the metathorax into the cavity of the propodeum (fig. 49 A), its only connections with the mesonotum being the attachments of its arms on the lateral ends of the scutellum. The intersegmental nature of this phragma, however, is shown by the fact that the great dorsal flight muscles of the mesothorax have their posterior attachments on it (figs. 48, 49 A, 71).

The pleuron of the mesothorax of the bee has little resemblance to that of a typical thoracic segment (fig. 29). It is the area below the mesonotum on the side of the thorax (fig. 30) between the pronotum in front and the second intersegmental groove (*2is*) behind, continuous below with the ventral sclerotization (S_2) before the middle leg. The upper part of the pleuron is traversed by an oblique groove (pls_2). Internally this groove forms a strong ridge (fig. 33 B, *PlR*), from the lower end of which there projects a short apodeme (*PlA*). These internal features identify the external groove as the pleural sulcus (A, *pls*), though its lower end runs into the intersegmental groove behind it and does not reach to the coxa. The part of the pleuron below the sulcus is therefore the episternum (*Eps*) and that above it the epimeron. The epimeron itself is divided by a groove (fig. 33 A) into an upper part (*Epm*) and a lower part (*epm*), but its extension to the coxa has been eliminated by the confluence of the pleural sulcus with the intersegmental groove. The upper part of the pleural ridge (fig. 33 B) joins a bridge, free from the pleural wall, that connects a large plate (*g*) inflected from the upper margin of the episternum with a smaller inflection of the epimeron (*Epm*). This complex endoskeletal structure apparently serves to strengthen the upper part of the mesopleuron.

The sternal area of the mesothorax is difficult to define, since the pleural and ventral sclerotizations of the segment are entirely confluent. The ventral surface of the mesothorax is marked by a median groove (fig. 33 C), from which is inflected the mesothoracic elements of a large endosternal structure (fig. 34, $Endst_{2+3}$) derived from both the mesothorax and the metathorax. It has been contended, therefore, that the episterna are actually continued, as they appear to be, to the midline of the venter and that the true sternum is infolded to form the endosternum. In the prothorax,

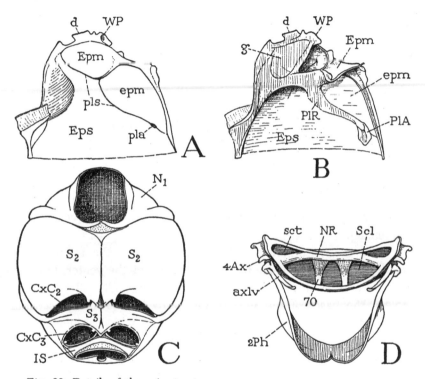

Fig. 33. Details of thoracic structure.

A, upper part of left mesothoracic pleuron. B, same part of right pleuron, inner surface. C, ventral surface of thorax, pleurosternal part of prothorax and legs removed. D, posterior notal plate of mesothorax and attached phragma, ventral.

d, basalar lobe of episternum; *g,* inflection of anterior margin of episternum to pleural ridge. For explanation of abbreviations see pages 92–93.

however, the large endosternum (fig. 31 B, *Endst*) is borne by a distinct sternal plate (D, *Fs*), and in the mesothorax the vertical muscles of flight, which in other insects are tergosternal muscles, are attached ventrally on the lower walls of the segment (fig. 48, 72) and not on the endosternum. It seems hardly likely that a receding sternum should pass its muscles on to usurping episterna; displaced skeletal parts usually take their muscles with them. Since the epidermis of the integument is continuous from one body region to another, there is no reason why the cuticular sclerotization may not become confluent where continuity gives some mechanical advantage.

The Metathorax— In the bee the metathorax is reduced to the narrow sclerites disposed in a crescent on each side wedged between the mesothorax and the propodeum (fig. 30). The metanotum (N_3) forms an arch over the back, widened downward on the sides and closely attached to the mesothoracic scutellum; it is not divided into a scutum and scutellum. Below the wing base on each side is a long pleural plate, divided into a larger upper part (Pl_3) and a small lower part (pl_3) adjacent to the leg base. The metapleuron retains no trace of a pleural sulcus and is hence not divided into an episternum and an epimeron. The coxa of the third leg is articulated laterally on a small condyle at the junction of the ventral pleural margin with the propodeum, from which fact it would appear that in the metathorax the epimeron has been entirely eliminated.

As in the mesothorax, so in the metathorax, the sclerotization of the pleural areas is continuous with that of the venter, but here the sternum is only a narrow transverse bridge (fig. 33 C, S_3) between the bases of the second and third legs. The metasternum likewise is divided by a median groove from which is inflected the metathoracic component of the endosternum.

The segmental parts of the composite endosternum of the wing-bearing segments (fig. 34, $Endst_{2+3}$) arise as inflections from the median grooves of their respective sterna. Each gives off a pair of divergent arms, which for a short distance unite on each side, and then proceed again separately. The longer anterior, mesothoracic arms (fig. 37 A, SA_2) curve laterally and upward to the pleural ridges, where they are attached to the pleura by short muscle fibers (79); their bases are united by a median bridge over the nerve cords. The shorter metathoracic arms (SA_3) go laterally and posteriorly over the bases of the second legs and unite with the ridges of the grooves separating the metapleura from the propodeum. The endosternum gives attachment to the longitudinal ventral body muscles, and to muscles of the second and third legs.

The Propodeum— The propodial segment of the thorax consists of a large dorsal plate and a narrow ventral plate. The dorsal plate,

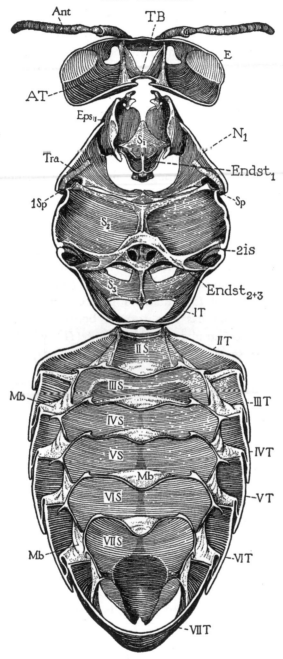

Fig. 34. Ventral skeleton of head, thorax, and abdomen, internal surface.

For explanation of abbreviations see pages 92–93.

or tergum (fig. 30, *IT*), is strongly convex and steeply declivous to the abdominal petiole. Its base is closely attached to the meta-thorax along the margins of the metanotum and the metapleura. On each side it contains a spiracle, which primarily is the first spiracle of the abdomen. The propodial sternum is a narrow plate arched behind the bases of the third legs (fig. 33 C, *IS*).

That the propodeum is derived from the abdomen is easily seen in the developing pupa, as has been shown by Zander (1910). In the bee larva (fig. 2 A) there is no constriction between the thorax and the abdomen. During the first part of the pupal period (E, G) the thoracic segments are marked by the presence of the newly everted legs and wings, but the first abdominal segment (*I*) still shows no particular change from the larval condition. At a later period (H), however, the first abdominal segment is crowded against the contracted metathorax (*3*); and finally, in the mature pupa (I) it has become reduced in size and so closely attached to the metathorax that it is virtually a part of the thorax. A deep secondary constriction has now formed between the propodeum and the primary second abdominal segment (*II*), which latter thus becomes the first segment of the adult abdomen.

Explanation of Abbreviations on Figures 29–34

Acx, antecoxal plate.

ANP, anterior notal wing process.

Ant, antenna.

51Ap, apodeme of muscle 51 (fig. 15).

aPh, prephragma.

AT, anterior tentorial arm.

atg, acrotergite.

4Ax, fourth axillary sclerite.

axlv, axillary lever.

Ba, basalare.

Bs, basisternum.

Cvx, neck, cervix.

Cx, coxa.

CxC, coxal cavity.

Endst, endosternum.

Epm, epm, subdivisions of epimeron.

Eps, episternum.

Fs, furcasternum.

HB, hypostomal bridge.

1is, 2is, 3is, first, second, and third intersegmental grooves of thorax.

L, leg.

Mb, intersegmental membrane.

N, notum.

NR, internal notal ridge.

ns, external notal sulcus.

92

pla, pit of pleural arm.

PlA, pleural arm.

Pl$_3$, *pl$_3$*, pleural plates of meta-thorax.

pls, pleural sulcus.

PN, postnotum.

PNP, posterior notal wing process.

pPh, postphragma.

S, sternum (*IS*, propodeal sternum, *IIS-VIIS*, abdominal sterna).

Sa, subalare.

Scl, scutellum.

Sct, scutum, *sct$_2$*, subdivisions of mesoscutum.

sf, scutal fissure.

Sp, spiracle (*1Sp*, first spiracle).

spl, spiracular lobe of pronotum.

T, tergum (*IT*, propodeal tergum; *IIT*, *VIIT*, abdominal terga).

TB, tentorial bridge.

Tg, tegula.

Tr, trochanter.

Tra, trachea.

W, wing.

WP, pleural wing process.

❈ VI ❈

THE LEGS

THE three pairs of legs of an insect are carried by the first, second, and third segments of the thorax. In the living insect the legs take such variable positions that for descriptive purposes we need a conventional system for designating their surfaces. Hence, regardless of the actual position of a limb, the surfaces are named according to the direction they face when the leg is extended straight out from the side of the body. In this position the upper surface is *dorsal,* the undersurface *ventral,* the preaxial surface *anterior,* and the postaxial surface *posterior.*

GENERAL STRUCTURE OF AN INSECT'S LEG

The leg of an insect is merely a slender, hollow, tubular outgrowth of the body wall provided with muscles, but by division into sections movable on one another it becomes a very efficient organ of locomotion. The "division" of the leg, however, is functional and not actual. The leg tube is simply differentiated into short unsclerotized parts interpolated between longer sclerotized parts. The unsclerotized parts are the *joints,* which permit of movement; the sclerotized parts are the *leg segments,* or *podites,* except that a segment may be subdivided by one or more secondary joints. For controlled movement at the joints, however, it is necessary that adjacent segments have specific points of contact with each other, and the hinging of the segments on each other is accomplished by the extension of articulating processes, or *condyles,* through the connecting joint membrane. There may be a single articulation between two segments, which is usually dorsal, and the joint is then said to be *monocondylic.* More commonly, how-

94

ever, a joint is *dicondylic,* having two articulations, which are either anterior and posterior, or dorsal and ventral. A monocondylic joint has much freedom of movement. At a dicondylic joint with a horizontal hinge the distal segment has only movements of *levation* and *depression,* termed also *extension* and *flexion.* If the hinge is in the vertical plane of the limb, the movements are *production* and *reduction.*

The basal segment of an appendage likewise is articulated on the body, in some cases by lateral and mesal articulations, in others by a single lateral articulation. If the limb as a whole swings forward and backward on the body, its movements are *promotion* and *remotion;* if it swings transversely, the movements are *abduction* and *adduction.* Since the various joints of a leg have different types of articulation and musculature, the leg as a whole is capable of varied movements. Leg muscles are named according to the movements they produce.

A typical, fully developed insect leg (fig. 36 A) has six segments, namely, a basal *coxa* (*Cx*), a *trochanter* (*Tr*), a *femur* (*Fm*), a *tibia* (*Tb*), a *tarsus* (*Tar*), and a *pretarsus* (*Ptar*). A leg *segment* may be defined as a section of the limb independently movable by muscles attached on its base, except where there is evidence of union between segments or of suppression of muscles. The **tarsus** may be a single segment, but it is usually divided into subsegments, or *tarsomeres,* which, though movable on each other, are not "tarsal segments," since they are not interconnected by muscles, the tarsal muscles being attached only on the basal tarsomere. The claw-bearing pretarsus is a true, independently musculated segment of the leg corresponding with the dactylopodite of a crustacean; its claws, therefore, are not "tarsal claws," as they are commonly termed in taxonomic descriptions.

DEVELOPMENT OF THE LEGS

In those insects that grow from embryo to adult by direct development, the legs appear on the embryo as three pairs of simple outgrowths on the thorax (fig. 10 B) and gradually develop into the limbs of the adult. In insects with a pupal stage, however, the larva, as in flies and bees (fig. 2 A), may have no apparent legs. In these insects the legs either do not appear at all in the embryo, or their rudiments disappear before hatching. In the embryo of the honey bee 60 to 62 hours old, according to Nelson (1915), the leg rudiments are clearly visible as low, rounded protuberances (fig. 10 C, *L*), but prior to hatching they become reduced to mere thickenings of the epidermis. In the larva, however, the leg tissue again begins to grow, but the leg buds now sink into pockets of the epidermis beneath the cuticle, so that practically, but not actually, the bee larva is legless. During the last larval stage, when the cuticle has been loosened preparatory to pupation, the leg buds evert from their pockets (fig. 2 C) and expand to a considerable length beneath the larval cuticle (E), finally (H) taking on the form of the pupal legs (I), which have approximately the size and shape of the legs of the adult bee. Within the cuticle of the pupal legs, the legs of the adult are now fully matured and at last are exposed at the shedding of the pupal cuticle. All these developmental processes are apparently for the purpose of relieving the larva of needless external legs and of restoring the legs in a functional condition to the adult.

THE LEGS OF THE ADULT HONEY BEE

The three pairs of legs of the adult bee are attached at somewhat different angles on the thorax so that in their action they are radially distributed from the sides of the body (fig. 3). In the worker the legs are modified for several purposes besides that of walking. The bee is an artisan of many trades and is provided with tools for all its needs, and some of its most important tools are parts of its legs. Each leg, however, preserves all the segments of a typical leg, with five subsegments in the tarsus. Inasmuch as the essential structure of the legs is the same in all three pairs, we may avoid much repetition by studying each segment separately rather than each leg as a unit. It will be convenient also to treat separately the two principal instruments of the legs which have nothing to do with locomotion, namely, the *antenna cleaner* and the *pollen press*.

The Coxa— The basal segments, or coxae, of the legs of the bee are articulated on the thorax between the pleura and the sternum of their segments, so that in general the legs turn forward and backward, but slight differences in the angles of the coxal axes give somewhat different movements to the three pairs of legs. The coxae of the forelegs hang downward on transverse axes between

96

their pleural and sternal articulations so that the first legs swing directly forward and backward on their bases. The coxae of the middle and hind legs are set obliquely on the thorax with their axes slanting somewhat posteriorly from the sternal to the pleural

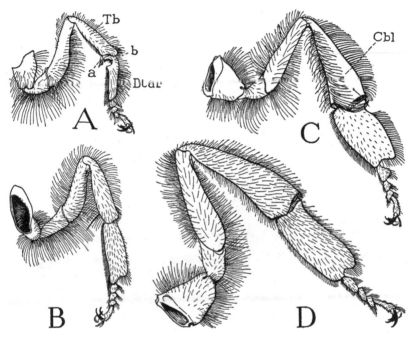

Fig. 35. Legs of a worker and hind leg of a queen.
A, left front leg of worker, anterior. B, left middle leg of worker, anterior.
C, left hind leg of worker, anterior (outer surface). D, hind leg of queen.

a, clasp (fibula) of antenna cleaner; *b,* notch of antenna cleaner. For explanation of abbreviations see page 112.

articulation (fig. 33 C) so that these legs turn outward with the forward movement and inward with the posterior movement. The hind legs are attached on the narrowed posterior part of the thorax and are usually directed backward.

The coxae are equipped with promotor and remotor muscles attached on their bases respectively before and behind their axes of movement (fig. 37 B). The coxal muscles of the bee arise on the endosternum and the pleura of their segments (A), except one in the prothorax that arises on the notum. In the more generalized orthopteroid insects, however, the principal coxal muscles take

97

their origins on the back plates of the segments. In each of the wing-bearing segments of the bee a slender muscle from the subalar sclerite is attached below on the coxa, but these muscles are functionally wing muscles.

The Trochanter— The short second segment of each leg, the trochanter (fig. 36 A, *Tr*), has a dicondylic articulation on the

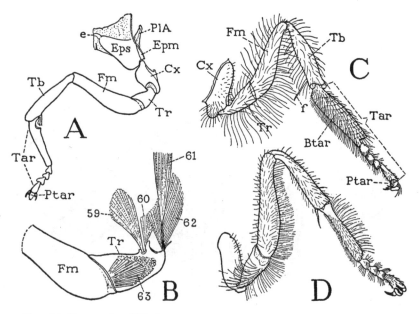

Fig. 36. Fore and middle legs.
A, left foreleg of worker suspended from pleuron of prothorax. B, base of right foreleg of drone, anterior. C, right middle leg of worker, posterior. D, right middle leg of drone, posterior.

e, pleural occipital process supporting head; *f,* tibial spine; *59, 60,* levator muscles of trochanter; *61, 62,* depressor muscles of trochanter; *63,* reductor muscle of femur. For explanation of abbreviations see page 112.

coxa with a horizontal transverse axis. The telopodite, or part of the leg beyond the coxa, therefore, turns directly up and down at the coxotrochanteral joint, and the trochanter is provided with levator and depressor muscles. The depression of the legs, however, is the force that sustains or lifts the body, or perhaps gives the insect a spring into the air preliminary to flight. The levator muscles of the trochanter (fig. 36 B, *59, 60*; fig. 37 C, *107, 108*) arise in the coxa and are inserted distal to the axis of the coxotrochanteral hinge. The necessarily stronger depressor musculature includes one

98

or two muscles from the coxa (fig. 36 B, *62;* fig. 37 C, *110*) and a larger muscle from the thorax, which in the prothorax arises on the pleural apophysis (fig. 15 C, *61*) and in the other two segments on the endosternum (fig. 37 A, *109*).

The trochanter is attached to the base of the femur on an oblique hinge with dorsal and ventral articulations (figs. 36 B, 37 C), which allows a slight forward and rearward movement to the femur on the trochanter but prevents any movement of the femur on the trochanter in a vertical plane. This device enables the trochanter to lift or depress as a unit the entire part of the leg beyond the coxotrochanteral hinge. Within the trochanter there is only a small reductor muscle of the femur (fig. 36 B, *63;* fig. 37 C, *111*).

The Femur— The elongate femur (fig. 36 A, *Fm*) is usually the principal segment of an insect's leg, but in the bee the more distal segments are disproportionately large. At the end of the femur is the prominent "knee" joint of the leg where the tibiotarsal section of the limb is extended or flexed on the trochanterofemoral section. The tibia turns on a horizontal axis between anterior and posterior articulations on the femur, and thus has strictly a hinge movement in a vertical plane. The whole length of the femur is occupied by the tibial muscles, which are a dorsal extensor (fig. 37 D, *112*) attached by a tendon to a dorsal process on the base of the tibia and a large ventral flexor (*113*) ending in a tendon attached to a small plate (*gf*) in the ventral membrane of the femorotibial joint. In addition to these muscles the femur contains a long branch of the flexor muscle of the pretarsus (*117a*), the tendon of which goes through the tarsus and tibia into the femur. There is also in the femur of each leg a fine tendonlike thread (*o*) extending proximally from its attachment on the head of the tibia just anterior to that of the extensor tendon. This tendon is probably the attachment of a chordotonal organ of the femur, such as occurs in the femur of various other insects.

The Tibia— In the fore and middle legs the tibia is a slender segment a little shorter than the femur (fig. 36 A, C, *Tb*). The tibia of the hind leg, however, is greatly elongated, flattened, and widened distally, which features are most pronounced in the worker (fig. 42 A, *Tb*), the hind tibia of the drone (F) and of the queen being more slender. The outer side of the tibia of the hind leg of the worker has a smooth and somewhat concave surface, which is bordered on both sides by a fringe of long, incurved hairs (fig. 42 C), forming the basket, or *corbicula* (*Cbl*), in which the workers transport pollen and propolis to the hive. The head of the tibia in all the legs is somewhat bent toward the femur, and the end of the lower wall of the femur is deeply emarginate to permit the flexion of the tibia against it.

The tibiotarsal joint differs from the other leg joints in being monocondylic. The single articulation between the tibia and the tar-

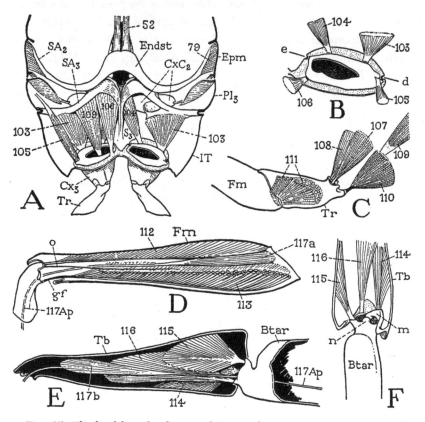

Fig. 37. The hind leg of a drone and its muscles.

A, lateroventral skeleton of posterior part of thorax and muscles of hind legs, dorsal. B, base of right hind coxa and muscles. C, trochanter and base of femur of right hind leg, with muscles, anterior. D, right hind femur and tibial muscles, anterior. E, right hind tibia and tarsal muscles, posterior. F, tibiotarsal joint of right hind leg, and muscle attachments, dorsal.

d-e, coxal axis of articulation; *m,* articular process of tibia; *n,* articular process of tarsus. Muscles are explained in text. For explanation of abbreviations see page 112.

100

sus is median and dorsal, and is essentially the same in all the legs, though somewhat modified in the hind legs. The tarsus is moved by three muscles arising in the tibia (fig. 37 E), two of which (*114, 115*) in the fore and middle legs are attached on the tarsus at the sides of the articular condyle and evidently serve to give anterior and posterior movements to the tarsus on the tibia. The third muscle (*116*) is attached in the ventral membrane of the joint and is a depressor of the tarsus. Besides the tarsal muscles the tibia contains a long branch (*117b*) of the retractor muscle of the pretarsus.

In the hind leg the articular process on the tarsus (fig. 37 F, *n*) curves downward, bringing the point of articulation almost to the center of the distal end of the tibia. As a consequence, the attachment of the posterior tarsal muscle (*115*) is so far above the articular point that this muscle in the hind leg acts as a strong levator of the tarsus on the tibia. The anterior muscle (*114*), however, is attached nearly opposite the articulation and retains its function of production of the tarsus. The ventral muscle (*116*), as in the other legs, is the tarsal depressor.

The Tarsus— The tarsus of each leg is divided into five subsegments, or *tarsomeres* (fig. 38 A, *Tar*), but the basal tarsomere is much longer than the others, and in the hind leg is broad and flat (fig. 42 A, *Btar*). This basal subsegment of the tarsus has been called the "metatarsus" and the "planta," but the first term should refer to the entire tarsus of the metathorax, and the second means literally the "sole of the foot." Neither term, therefore, is appropriate; *basitarsus* is preferable. The long, cylindrical, bristly basitarsus of the first leg (fig. 35 A) is used as a brush for removing pollen or other particles from the head and fore parts of the body, but the most important feature of the first leg is the *antenna cleaner* at the base of the tarsus.

The basitarsus of the hind leg is equally large in each caste (fig. 42 A, F), but there is no apparent reason for its size and shape in the queen and drone. In the worker (A), however, the broad posterior (inner) surface of the basitarsus is armed with nine transverse rows of long stiff spines, directed distally at an angle of 45 degrees to the surface, and thus resembles a flat brush. It serves for the collection and retention of pollen to be stored in the pollen basket of the tibia. The deep notch in the dorsal margin of the hind

101

leg between the tibia and the basitarsus is converted in the worker into a *pollen press* (A, *Pr*) for transferring pollen from the basitarsal brush of one leg to the tibial basket of the other, as will be explained in the last section of this chapter.

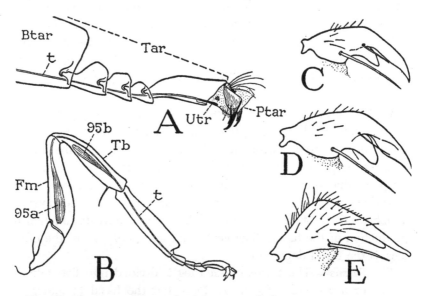

Fig. 38. The tarsus and muscle of the pretarsal claws.
A, right hind tarsus of drone, posterior. B, diagram of leg showing distribution of pretarsal muscle. C, anterior claw of hind leg of worker, anterior. D, same of queen. E, same of drone.

t, tendon of pretarsal muscle; *95*, pretarsal muscle. For explanation of abbreviations see page 112.

The four small subsegments of the tarsus (fig. 38 A) are freely movable on each other by monocondylic articulations, as is the basitarsus on the tibia, but they contain no muscles to give them individual movement. The entire tarsus, however, is traversed by the tendon of the flexor muscle of the pretarsus (A, B, *t*), and a pull on this tendon causes a deflection of the slender part of the tarsus on the basitarsus.

The Pretarsus— The terminal segment of the insect leg, for want of a better name, is termed the *pretarsus*. Functionally it is the foot of the insect, and, whatever it may be called, it must be understood that it is a true segment of the leg corresponding with the dactylopodite of a crustacean. The pretarsus is articulated on the end of the tarsus and is movable by its own muscles. In the spiders and crustaceans the end segment of the leg has two muscles, a levator and a depressor, arising in the tarsus; but the insects and myriapods have dispensed with the dorsal muscle, and the ventral depressor has come to arise in the tibia and femur and is commonly termed the *retractor of the claws*. The fibers of the muscle are attached on a long tendon from a plate at the base of the pretarsus (fig. 38 A, *Utr*), which extends through the tarsus and tibia into the femur (B).

The pretarsus of the honey bee has an unusually complex and specialized structure (fig. 39). It consists essentially of a median part, which is the apical segment of the leg, projecting from the end

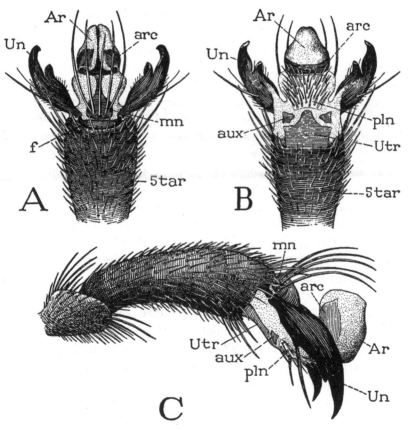

Fig. 39. Last tarsomere and pretarsus of a worker. A, dorsal; B, ventral; C, lateral.

f, marginal flange of fifth tarsomere. For explanation of abbreviations see page 112.

103

of the fifth segment of the tarsus (A, B, *5tar*), and of a pair of lateral claws, or *ungues* (*Un*). The claws arise from the base of the segment (B), but they are articulated individually on condyles borne on a small dorsal plate (A, *f*) at the end of the tarsus. The body of the pretarsus ends with a soft apical lobe, the *arolium* (*Ar*), ordinarily turned upward between the claws (C), which serves as an adhesive organ when the bee is on a surface too hard or smooth for the claws to grasp.

The arolium in the passive position presents dorsally a deep cavity between its upturned lateral walls (fig. 40 A, *Ar*). In its outer wall is a dark, elastic, U-shaped band, the *arcus* (B, E, *arc*), which embraces the base of the arolium ventrally with its arms extending distally in the lateral walls (F). Otherwise the outer walls of the arolium are soft, smooth, and flexible.

The basal part of the pretarsus contains dorsally an elongate, flask-shaped, median sclerite (fig. 40 A, E, *mn*) armed with five or six long, thick, curved spines. This sclerite is articulated proximally on the end of the tarsus between the bases of the claws, and by its narrowed distal end it is attached like a handle to the base of the arolium, for which reason it may be termed the *aroliar manubrium*. Ventrally, just proximal to the arolium, is a broad, weakly sclerotized plate, the *planta* (B, E, *pln*), thickly beset with strong spines diverging distally. Proximal to the planta is a second, larger, shield-shaped plate, which is the *unguitractor* (*Utr*). The unguitractor is more or less retracted into a membranous infolding of the distal end of the tarsus, and close to its base arises the long apodemal tendon of the retractor muscle (*95t*). At the distal angles of the unguitractor are two small *auxiliary sclerites* (*aux*).

The claws are hollow, strongly sclerotized outgrowths of the membranous lateral walls of the pretarsal base, from which each claw arises between its articular condyle above and the auxiliary sclerite below (fig. 40 E). The claws are thus freely flexible on the pretarsus, but they themselves are devoid of muscles. Each claw has two points of unequal length (C, D) and two long spines on the outer surface. The claws of the worker (fig. 38 C) and the queen (D) are similar in shape, but those of the drone (E) are angularly bent and have long, slender points.

The pretarsal muscle, as already noted, includes a group of fibers arising in the tibia and a larger group with its origin in the femur, both attached on branches of the unguitractor tendon. The contraction of the muscle pulls the unguitractor plate into the end of the tarsus, and the tension is exerted on the bases of the claws through the auxiliary sclerites. The result is that the claws are flexed on their tarsal articulations and hold by their points on the supporting surface, unless the latter is too smooth for them to grasp.

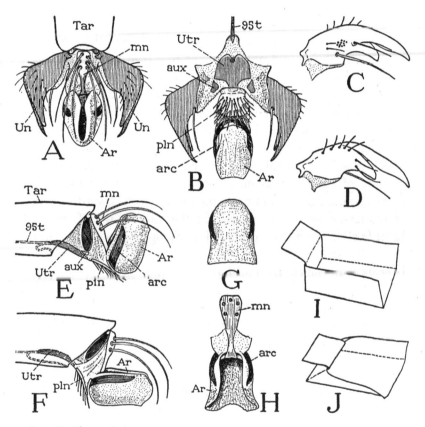

Fig. 40. The pretarsus.

A, pretarsus of middle leg of drone, dorsal, flattened, claws spread, arolium turned up. B, same, ventral, arolium depressed and spread out. C, posterior claw of middle leg of worker, outer surface. D, anterior claw of hind leg of worker, inner surface. E, pretarsus and end of tarsus, claws removed. F, same, arolium depressed. G, ventral surface of flattened arolium. H, arolium and manubrium, dorsal. I, paper model of arolium and manubrium in nonfunctional position. J, same, illustrating flattening of arolium by downward pressure on manubrium.

95t, tendon of pretarsal muscle. For explanation of abbreviations see page 112.

In the latter case the claws sprawl helplessly forward with their points directed outward. The continuing pull of the muscle exerted on the base of the pretarsus now flexes the pretarsus itself and brings the arolium down into a horizontal position (fig. 40 F), after which the arolium unfolds, spreads out flat, and clings to the smooth surface of the support on which the claws cannot hold. Thus the bee is provided with means for keeping its foothold on whatever kind of surface it may find itself.

The mechanism of the extension and flattening of the arolium can be demonstrated in a dead bee. If the foot is in the position shown at F of figure 40, an upward pressure on the retracted planta (*pln*) extends the pretarsus until the planta comes into the same horizontal position as the arolium, and now, with continued pressure, the arolium itself automatically unfolds and spreads out with its flattened undersurface downward. There can be little doubt, therefore, that in the live bee this same action results from the downward pressure of the tarsus against the foot. With the extension of the pretarsus, the manubrium turns posteriorly and upward on the end of the tarsus and hence pushes on the upper edge of the aroliar base; finally, when fully extended, it clamps down on the arolium, which is thereby expanded and securely held against the surface beneath it in the spread position. The action of the arolium can be illustrated with a piece of paper cut and folded into the form of a scoop (I). A downward pressure on the base of the scoop spreads the sides widely apart (J), giving a good imitation of the partly expanded arolium (H).

The spreading action of the arolium, therefore, results automatically from the pull of the retractor muscle on the pretarsus and the downward pressure of the tarsus on the base of the pretarsus when the claws fail to grasp the support. On release of the pressure the elastic arcus in the aroliar wall once more folds the aroliar walls upward in the nonfunctional position, and the claws evidently are extended by elasticity of the pretarsal base on relaxation of the retractor muscle. It has been said that the arolium adheres to smooth surfaces by means of a sticky exudation from the spines of the planta, but observation fails to reveal the presence of any such adhesive; a bee walking on glass leaves no smear. The upper surface of the arolium is covered with minute hairs, but its undersurface is almost entirely bare.

THE ANTENNA CLEANER

Insects are meticulous about keeping their antennae free of dirt, since the antennae are the seat of important sense organs. Species having biting mouth parts and filamentous antennae clean the latter by drawing them through the lobes of the maxillae. Others clean the antennae with the front legs, which may be equipped with hairs or spines for the purpose, but the Hymenoptera have on the front legs a special gadget for cleaning the antennae, which is best developed in the bees.

The antenna cleaner of the worker honey bee (fig. 41 B) consists

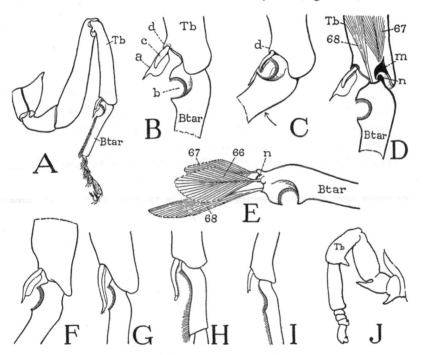

Fig. 41. The antenna cleaner.
A, left foreleg of worker, showing antenna cleaner at base of tarsus. B, antenna cleaner open. C, same, closed. D, tibiotarsal joint of front leg and tarsal muscles in tibia. E, base of first tarsomere with attached muscles, anterior. F, antenna cleaner of *Halictoides calochorti* Ckll. G, same of *Vespula maculata* (L.). H, same of *Trogus vulpinus* (Grav.). I, same of *Orussus sayi* Westw. J, foreleg of honey-bee pupa, showing tibial spur.
a, closing lobe (fibula) of antenna cleaner; *b,* notch of antenna cleaner; *c,* anterior lobe of fibula; *d,* tibial process at base of fibula; *m,* articular process of tibia; *n,* articular process of basitarsus; *66, 67,* extensor muscles of tarsus; *68,* flexor muscle of tarsus. For explanaton of abbreviations see page 112.

of a deep semicircular notch (b) on the inner (ventral) surface of the proximal end of the basitarsus ($Btar$) and of a large, flattened spur (a) projecting from the inner angle of the distal end of the tibia (Tb) which closes like a clasp, or *fibula*, over the tarsal notch. The concavity of the notch is armed with a comb of fine, closely set, spinelike hairs. The fibula is a broad, thin, movable appendage with a narrowed base, a sharp distal point, and a strong, spatulate accessory lobe (c) on its anterior surface. The fibula has no muscles, but when the basitarsus is flexed on the tibia (C), the notch on its base is brought against the fibula and the notch is thus closed to a circular aperture. The fibula resists the pressure of the tarsus against it by the point (d) that projects above its base from the end of the tibia. The basitarsus, having a single, dorsal point of articulation with the tibia (D, n), is freely movable on the latter, and its three muscles (E) are attached on three sides of the articular condyle.

When the bee uses the antenna cleaner, by appropriate movements of the leg the tarsal notch is first placed around the base of the antennal flagellum and then, by flexion of the tarsus, the antenna is brought against the fibula and thus securely held in the cleaner. The antenna is now drawn upward; the tarsal comb cleans its sensory outer surface; the thin accessory lobe of the fibula scrapes its inner surface. The antennal cleaner is present in the drone and queen as in the worker and is developed to some degree in most of the Hymenoptera (fig. 41 F–I), though the tarsal notch is usually shallower than in the honey bee and may be but a sinuosity of the tarsal margin (I). The fibula appears to be merely a modified spine on the end of the tibia; in the honey-bee pupa (J) it is represented by a short tibial spur.

THE LEGS AS CARRIERS OF POLLEN AND PROPOLIS

Pollen forms an important part of the food of the bees, both of the adults and the larvae. Though it is commonly called "bee bread," it might better be said to be the "meat" of the bees, since it is the principal source of protein in their diet. The foraging bees, therefore, collect pollen as well as nectar. Propolis is the resinous gum that bees obtain mostly from the buds of trees and use for stopping cracks or crevices in the hive, and also for strengthening the wax in comb building. Both pollen and propolis are carried to the hive in the baskets of the hind tibiae.

In their visits to flowers for nectar the bees unavoidably become dusted all over with pollen, but when they are foraging specifically for pollen, according to Parker (1926), they deliberately dislodge the pollen from the anthers. The pollen falling on the anterior part of the body is moistened with regurgitated honey and thus becomes sticky and adherent. From the head and fore parts of the body the pollen grains are cleaned off by the pollen brushes of the front legs, and from the latter they are scraped off upon the broad inner surfaces of the basitarsi of the middle legs. These legs are then grasped one at a time between the brushes of the basitarsi of the hind legs and drawn forward, thus transferring the pollen to the inner (posterior) surfaces of the hind basitarsi. When the basitarsal brushes are sufficiently loaded, the pollen is finally stored in the tibial baskets. The ability of the bee to make this transfer, however, depends on the instrument known as the pollen press, developed in the worker between the end of the tibia and the basitarsus.

The structure and the action of the pollen press have been described by several writers, including Sladen (1911), Casteel (1912b), Beling (1931), and others. In the hind leg of the queen and drone as well as in the worker there is a deep, pincerlike dorsal notch between the adjacent ends of the tibia and the basitarsus (fig. 42 F). It is this notch that is elaborated in the worker to form the pollen press (A, Pr). The tarsal lip of the notch in the worker is widened and beveled outward and upward toward the tibia and is expanded in a small lobe known as the *auricle* (B, au). The auricle is bordered by a fringe of long hairs; its surface is covered with small spicules and is limited mesally (posteriorly) by a ridge on the end of the basitarsus. The opposing tibial lip of the notch is armed on its mesal margin by a row of closely set spines (ras), forming a comb (*pecten*), or little rake (*rastellum*), which projects down over the base of the auricle (A). When the basitarsus is turned upward at the tibiotarsal joint, the fringed margin of the auricle passes within a semicircle of long curved hairs on the lower end of the outer face of the tibia (D) and overlaps the lower end of the floor of the pollen basket (E).

Preparatory to operating the pollen presses for loading the pollen baskets, the bee brings her hind legs together and moves them alternately up and down. By this action the rastellum of the descending legs scrapes a small mass of pollen from the basitarsal

109

brush of the other leg. The pollen detached by the rastellum falls on the surface of the auricle and adheres to it. Then by an upward flexion of the basitarsus on the tibia, the pollen on the auricle, held in place by the rastellum, is pressed up into the lower end of the tibial

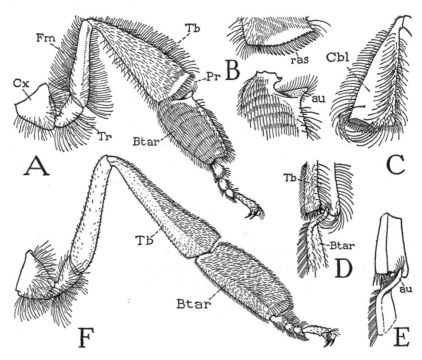

Fig. 42. Hind leg of worker and drone and pollen press of worker.
A, right hind leg of worker, posterior (inner) surface. B, ends of tibia and basitarsus of hind leg, separated, posterior. C, pollen basket (corbicula) on outer surface of hind tibia of a worker. D, pollen press between tibia and tarsus of hind leg of worker, dorsal. E, same with tibial hairs removed. F, right hind leg of drone, posterior.
For explanation of abbreviations see page 112.

basket. With repeated scraping, first from one basitarsal brush and then from the other, and operation of the presses, the baskets are finally loaded by successive additions of pollen pressed upward from below. During most of this process of loading the pollen baskets, Parker says, the bee remains on the wing, often hovering in the air without any forward movement.

Though the tibiotarsal joint of the hind leg, as already explained, differs somewhat from that of the other legs, and one of its muscles acquires a strong levator action on the tarsus, the mechanism is the same in the queen and the drone as in the worker and is hence not correlated with the pollen press of the worker. However, the special structure at this joint is unquestionably conducive to the effective action of the press.

When the pollen-collecting bee has loaded both pollen baskets, she returns to the hive and goes to a cell of the comb in which pollen is to be stored. Here, according to Casteel (1912a), she rests her front legs on one edge of the cell, places the end of her arched abdomen against the other, stretches the pollen-laden hind legs into the cell, and pries their loads off with the basitarsi of the middle legs. The pollen falls to the floor of the cell, and the bee usually goes off without giving it further attention. Another bee then comes to the cell, breaks up the pollen mass with her mandibles, and tamps it down into the bottom of the cell.

Propolis also is carried to the hive in the baskets, or corbiculae, of the hind legs, but the method of collecting it and of loading the baskets is quite different from that used in the case of pollen. A foraging bee collecting propolis, as described by Rösch (1927b), bites off with her mandibles assisted by her forelegs a small piece of the gummy exudation from a bud of some tree. The particle is first kneaded in the jaws and then taken onto one of the forelegs. At the same time the hind leg of the same side is brought forward, and now the particle of propolis is scraped off the foreleg by the middle leg and quickly pressed into the corbicula of the hind leg. The action is then repeated on the other side of the body, while the bee supports herself on the legs of the opposite side. When the loaded bee arrives in the hive, she goes to some place where propolis has been used and there quietly waits for another bee to unburden her. The relieving bee bites off a small piece of propolis and then, carrying it in her mandibles, runs in a direct course to a place in need of cementing, where with her mandibles she presses it into the crevice.

Explanation of Abbreviations on Figures 35–42

Ar, arolium.

arc, arcus.

au, auricle.

aux, auxilia.

Btar, basitarsus.

Cbl, pollen basket, corbicula.

Cx, coxa.

CxC, coxal cavity.

Endst, endosternum.

Epm, epimeron.

Eps, episternum.

Fm, femur.

gf, genuflexor sclerite.

mn, manubrium.

Pl, pleuron.

PlA, pleural arm.

pln, planta.

Pr, pollen press.

Ptar, pretarsus.

ras, rastellum, rake.

S, sternum.

SA, arm of endosternum.

IT, tergum of propodeum.

Tar, tarsus.

5tar, fifth tarsomere.

Tb, tibia.

Tr, trochanter.

Un, unguis, lateral claw of pretarsus.

Utr, unguitractor.

THE WINGS

THE wings of all flying animals other than insects, including the modern bats and birds and the extinct flying reptiles, are simply modified front legs, either supporting folds of skin or provided with feathers. The wings of insects, on the other hand, are new structures evolved from folds of the integument, paranotal lobes, extended from the sides of the back on the second and third thoracic segments. The insects are thus not only furnished with organs of flight, but they retain intact their original equipment of legs. In this respect insects resemble the winged dragons and other such animals of fiction, though these creatures appear to have had no anatomical provision for moving their wings. Herein the insects were more fortunate. Their muscles being inside instead of outside the skeleton, and the skeleton itself being flexible and elastic, the muscles could vibrate the wing-bearing plates of the skeleton and thus give movement to the wings. Because of this favorable circumstance due to their structure, the insects have been able to evolve a wing mechanism that has made them unsurpassed flyers.

It is not to be supposed that those early insects that first acquired paranotal lobes on the mesothorax and metathorax had any preformed mechanism for flapping the lobes. Probably the lobes served primarily as gliders. If, however, the lobes became flexible at their bases, it was a simple matter, evolutionarily speaking, to draw the thoracic segments together so that contraction of the intersegmental muscles of the back would arch the notal plates upward and thus deflect the wing lobes on pleural fulcra supporting them from below. Conversely, a contraction of the pre-existing dorsoventral muscles could flatten the notal plates and elevate the wings. By this simple pump-handle mechanism the principal wing movements are still produced in most insects, but the activating muscles have been much enlarged and adapted to their new function.

A mere up-and-down flapping of winglike lobes, however, would not produce true flight, so other minor adjustments in the thoracic skeleton came about that gave the wings, along with their up-and-down movement, a torsion on their long axes, which converted them into organs for progressive movement in the air. This turning movement of the wings (fig. 44 C) that deflects the wing anteriorly with the downstroke and posteriorly with the upstroke is produced by the muscles of the epipleural sclerites of the thorax, which pull directly on these sclerites and secondarily on the wing bases alternately before and behind the pleural fulcra.

Finally, in addition to the movements of flight, the wings of most insects are capable of being flexed backward in a flat position over the body when not in use and then again extended to the flight position. The mechanism of flexion and extension depends on the mechanical relations of small *axillary sclerites* developed in the wing bases. Flexion of the wing is produced by a special flexor muscle attached on one of the axillaries, extension results from the pull of the basalar muscles.

DEVELOPMENT OF THE WINGS

Since the wings are a relatively late acquisition in the evolution of insects, they do not begin their ontogenetic development in the embryo. Their origin from padlike lobes extended from the sides of the back is well shown in those insects that develop gradually from the young to the adult. In such insects the wings arise in an early nymphal stage as small, flat, hollow outgrowths of the lateral margins of the back plates of the mesothorax and metathorax and, as they grow, take on the form and structure of mature wings. Each wing pad is penetrated by branches of the thoracic tracheae. Later the two surfaces of the pad come together and obliterate the lumen of the wing except for channels along the tracheae. The walls of the channels then become sclerotized and form the so-called *veins* of the wing, while the intervening areas take on a thin parchment-like structure. The wing pad thus develops into a membranous wing in which the principal veins are sclerotized tubes originally formed about the tracheae.

114

With insects that hatch as larvae and undergo a pupal metamorphosis into the adult, the wings first appear also in a postembryonic stage, but they begin their growth beneath the larval cuticle, usually in pockets of the epidermis, and do not normally appear as external appendages until the last larval skin is shed. In the bee larva the wing rudiments are contained in pockets of the epidermis beneath the cuticle, situated well *below* the spiracles of their segments (figs. 6 C, 84 A, W_2, W_3). When the wings are everted in the early propupal stage, however, they have the form of elongate flaps (fig. 2 G, W_2, W_3), and their bases are on a level with the spiracles. Finally, in later stages (H, I) the wings take their definitive position on the side of the back above the spiracles. Since in most holometabolous larvae, as in ametabolous nymphs, the young wing buds lie above the spiracles in the position of the adult wings, their ventral origin in the bee larva probably has no phylogenetic significance, but it is difficult to explain. In some of the holometabolous insects, particularly in the Hymenoptera, the wing veins are laid out first as thickenings of the wing surfaces, and their channels are later penetrated by tracheae. In this case the tracheae give no clue to the identity of the veins. The wings complete their development beneath the cuticle of the pupa, and when the adult comes out they have only to unfold, expand, and take on the mature form.

THE WING STRUCTURE

The wings of the honey bee (fig. 43) are greatly simplified in both their shape and in the pattern of their venation as compared with the wings of generalized insects. The forewing is much larger than the hind wing, but in flight the two wings of each side work together, and to ensure their unity of action the hind wing is linked to the forewing. The coupling apparatus consists of a row of minute hooks on the anterior margin of the hind wing (B, h) and a narrow fold on the posterior margin of the forewing (A, f). The hooks of the hind wing curve upward and backward from the wing margin, with a twist in the direction of the wing tip, and differ somewhat in shape in the worker (C) and the drone (D). When the wings are at rest and folded over the back, there is no connection between them, but when they are extended preparatory to flight, the front wing is drawn over the upper surface of the hind wing beneath it, and the hooks of the latter engage in the fold of the front wing (E).

The so-called *veins* of the wing are supporting ribs, and not literally veins, though blood flows along their channels. Their cavities, being remnants of the body cavity extended into the primitive wing lobes, allow the penetration of nerves into the wing and the circulation of blood through them.

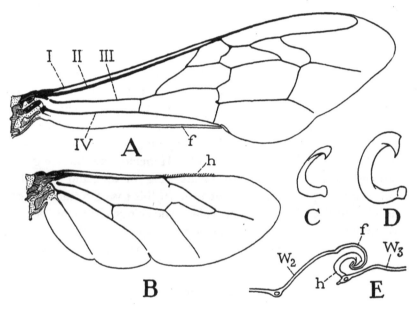

Fig. 43. The wings.

A, B, fore and hind wing of a drone. C, hook of hind wing of a worker. D, hook of hind wing of a drone. E, the interlocked wing margins.

f, fold on posterior margin of forewing; *h*, hooks on anterior margin of hind wing; *I-IV*, main veins of wing. For explanation of abbreviations see page 133.

The wings arise from the sides of the thorax between the notal and pleural plates of their segments. Each wing is articulated on the edge of the notum and is supported from below on the wing process of the pleuron (fig. 44 A, *WP*). The pleural fulcra lie a short distance laterad of the notal hinges, so that alternating downward and upward movements of the notum (A, B) turn the wings respectively up and down on the pleural supports. The flexible basal part of the wing is a two-layered membrane, the dorsal layer being continuous with the back, the ventral layer reflected into the pleuron.

Within the basal membranous area are several small sclerites, the *axillaries,* which play an important part in controlling the wing movements. The axillary mechanism of the bee is so specialized that we will understand it better by first examining the axillaries and their interrelationships in a more generalized type of wing.

Typically there are three principal axillary sclerites in the base of each wing (fig. 45), distinguished as the first, second, and third.

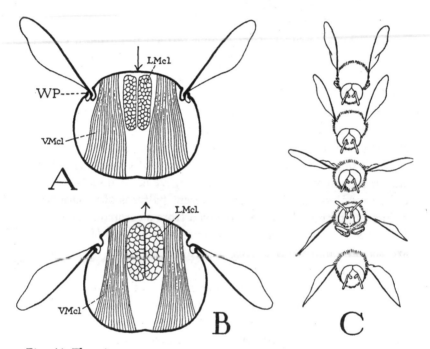

Fig. 44. The wing movements.

A, the wings turned upward on the pleural wing processes (*WP*) by depression of the notum caused by contraction of the vertical indirect wing muscles (*VMcl*). B, wings turned down by upward curvature of the notum produced by contraction of the longitudinal indirect wing muscles (*LMcl*). C, successive positions of the wings of a drone in flight (from photographs by Stellwaag, 1910).

The *first axillary* (*1Ax*) articulates on the anterior wing process of the notum (*ANP*); the *second axillary* (*2Ax*) is extended ventrally through the wing and rests on the pleural wing process, being thus the pivotal sclerite of the wing base; the *third axillary* (*3Ax*) usually articulates directly with the posterior notal wing process (*PNP*) and is functionally the flexor sclerite of the wing, being the only

117

axillary on which a muscle is attached. In some insects, as in the bee, a small *fourth axillary* (*4Ax*) intervenes between the third axillary and the posterior notal wing process. In addition to the axillaries, there is a *median* sclerite (*m*), or two such sclerites, lying between the second axillary and the bases of the median and cubital veins. At the anterior angle of the wing base is usually a small, hairy pad (*Tg*), which in the mesothorax of the bee is expanded into a large scalelike lobe, the *tegula,* overlapping the base of the front wing (fig. 30, *Tg*). The free posterior part of the axillary membrane is bordered by a corrugated thickening termed the *axillary cord* (fig. 45, *AxC*), which is continuous from the posterior margin of the scutellum into the posterior margin of the wing.

In a study of the wings of the honey bee we need give but small attention to the wing venation, but the axillaries are of much importance, and the epipleurites below the wing bases must be included as essential elements of the wing mechanism.

The Wing Venation— An important part of entomological taxonomy is a comparative study of the wing veins of insects, since it is to be supposed that the various different patterns of venation in the orders and families have been evolved from a common primary plan. In a theoretically generalized scheme of venation (fig. 45), there are six principal veins arising separately from the wing base and branching in a definite manner in the distal field of the wing. Behind the sixth vein there is usually a fold of the wing (*vf*) setting off a posterior lobe, or *vannus,* containing a variable number of veins radiating like the ribs of a fan from a common base. According to the well-known Comstock-Needham system of naming the wing veins, the first vein is the *costa* (*C*), the second the *subcosta* (*Sc*), the third the *radius* (*R*), the fourth the *media* (*M*), the fifth the *cubitus* (*Cu*), and all the others *anal veins.* Since the "first anal," however, has a separate origin and lies before the vannal fold (*vf*), it may better be termed a *postcubitus* (*Pcu*). The other "anals" in the fan region behind the fold, then, are appropriately called the *vannal veins* (*V*).

The wings of most insects conform pretty well with this plan of venation, but in each wing of the honey bee (fig. 43) there are only four main veins in all, and these break up into such an irregular pattern of secondary veins and cross veins that it is difficult to discover any relation between them and the veins of the generalized wing

118

(fig. 45). In some other Hymenoptera the wing venation is even more simplified than it is in the bees. Attempts have been made to apply the Comstock-Needham nomenclature to the veins of the hymenopterous wing, but the results have been unconvincing because of the difficulty of identifying the veins and their branches with those of other insects. As Ross (1936) has said: "Few problems

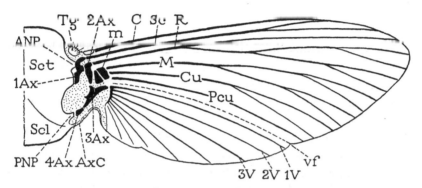

Fig. 45. Diagram of a generalized insect wing and its articulation on the notum.

m, median plate of wing base. For explanation of abbreviations see page 133.

in insect wing venation have brought forth such a mass of contradictory explanations and opinions as the question of the interpretation of the hymenopterous wing." Most specialists in Hymenoptera, therefore, have adopted a nomenclatural system of their own, which is different for different families. Since in the following discussion of the wings of the bee we shall be concerned with structural facts and the wing mechanism, rather than with the troubles of taxonomists, we may simply give a Roman numeral to each of the four veins arising from the wing base (fig. 43 A) and pass over the question of their homology with the veins of other insects.

The Axillaries of the Front Wings— In the naturally extended position of the front wing of the bee, the wing base is rolled on itself (fig. 46 A), the posterior part being turned downward so that little of it is visible when viewed from above. When the basal region is artificially spread out flat (D), however, it is seen that its structure closely conforms with that in the general plan given in figure 45. At the anterior end of the wing base is a large irregular sclerotization, apparently formed by the enlarged and united bases of the

119

first three veins, which may be termed the *humeral complex*. Lying obliquely just behind the humeral complex is an elongate *median plate* (*m*), and distributed in the membrane of the wing base are four distinct *axillary sclerites* (*1Ax, 2Ax, 3Ax, 4Ax*). The axillaries are quite different in size and shape, but the details of their structure are all of importance in the wing mechanism; the sclerites not only hinge the wing to the thorax, but they are essential mechanical elements in the control of the wing movements.

The *first axillary* (fig. 46 E, *1Ax*), is a large sclerite with an elongate, posteriorly tapering body and a large head (*b*) turned laterally on a curved neck, posterior to which is a large lateral projection (*f*). The first axillary is articulated to the margin of the mesonotum in such a manner that the neck rests on the anterior end of the anterior notal wing process, the body is hinged by an articular lobe (*c*) to the posterior end of the wing process, and the tapering posterior end (*g*) rests and turns in a concavity on the lateral margin of the posterior scutal plate of the mesonotum (J, *sct*). In the natural position of the extended wing (A) the first axillary stands almost vertically, but in the flattened wing base (D) it is seen that its head abuts against the humeral complex of the wing and that the lateral projection underlies the inner edge of the median plate (*m*). The first axillary is the anterior hinge plate of the wing.

The *second axillary* (fig. 46 E, *2Ax*) is a relatively small sclerite standing vertically in the wing base. Its exposed upper surface (*i*) appears as a small oval plate (D, *2Ax*) lying in the angle between the head and the body of the first axillary. Ventrally the second axillary articulates by an anterior knob (E, *j*) with the wing process of the pleuron (B, *WP*), and by a posterior arm (*k*) it is associated with the subalar sclerite (*Sa*) on the upper edge of the pleuron. The second axillary is the *pivotal sclerite* of the wing base, since it is the only axillary sclerite that rests on the pleuron and gives the wing a solid support from below. In the bee the second axillary has no specific association with any of the wing veins, but in most insects it is connected with the base of the radius (fig. 45, *2Ax*).

The *third axillary* (fig. 46 E, *3Ax*) is an elongate sclerite lying close against the axillary cord (D), with its tapering distal end closely connected with the enlarged base of the fourth vein of the wing and its proximal end associated with the fourth axillary. On its anterior margin the third axillary bears a large lobe (E, *l*) on

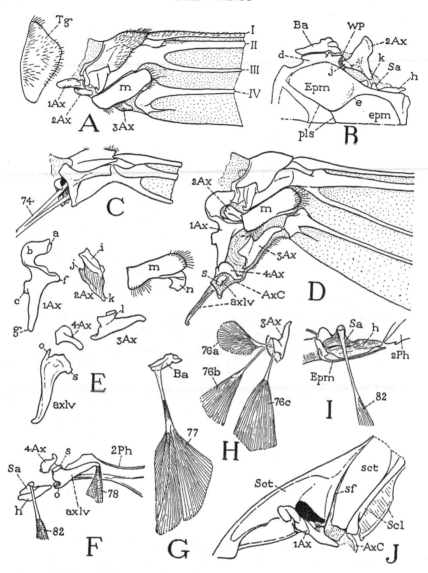

Fig. 46. Base of forewing of a drone, its articulations and its pleural muscles. A, base of forewing extended in natural position. B, pleural articulation of basalare and second axillary. C, humeral area of wing base, ventral. D, wing base, axillary area flattened, dorsal. E, axillaries of right wing, dorsal. F, fourth axillary and axillary lever. G, right basalare and muscle, mesal. H, right third axillary and muscles, mesal. I, right subalare and muscle, mesal. J, mesonotum with first axillary in natural position, lateral.

Alphabetical lettering and muscles are explained in text. For explanation of abbreviations see page 133.

which are inserted three muscles arising on the inner surface of the mesopleuron (H). The third axillary is the only sclerite of the wing base provided with muscles; it is the effective skeletal element of the wing-flexing mechanism and may be termed the *flexor sclerite*.

The *fourth axillary* (fig. 46 E, *4Ax*) is a small, irregularly triangular sclerite lying transversely in the posterior angle of the wing base (D), where it overlaps the posterior end of the first axillary and articulates with the posterior wing process of the scutum. Its narrow outer end is closely pressed against the axillary cord behind the proximal end of the third axillary. The fourth axillary is the *posterior hinge plate* of the wing in the bee, but in most insects a fourth axillary is absent, and the third axillary articulates directly with the posterior wing process of the notum. In bees, however, the fourth axillary has an accessory function in connection with the flexion of the wing because of its close association with the exposed end of an internal leverlike sclerite (D, *axlv*) connected with the arm of the postphragma (F, *2Ph*).

The *axillary lever* of the honey bee is an elongate, triangular sclerite (fig. 46 E, *axlv*) lying close against the mesal surface of the arm of the postphragma (fig. 33 D). The only exposed surface of the lever is the dorsal angle of its broad base (fig. 46 F, *s*), which appears as a small nodule in the wing membrane immediately behind the fourth axillary (D, *s*). By a point on the lower angle of its base (F, *o*) the lever articulates with the arm of the phragma (*2Ph*). On its tapering distal end is attached a muscle (78) from the mesothoracic arm of the endosternum. The function of the lever will be explained in connection with the mechanism of the wing.

So far as observed, the axillary lever is a free sclerite only in the Apoidea. In other Hymenoptera it is represented by an immovable process or lobe of the phragma arm, on which, however, a muscle from the mesothoracic endosternum is always attached. In wasps the process is so similar in shape to the free lever of the bee that there can be no question of the homology of the two structures, but in other families it is a simple muscle-bearing projection from the lateral part of the phragma. The development of this musculated process of the phragma into a movable accessory of the wing-flexing apparatus in the bees is a highly interesting example of the evolution of a special mechanism from a structure having originally a quite different function.

The Epipleurites of the Mesothorax— The epipleurites include in each wing-bearing segment the *basalare* resting on the upper edge of the pleuron before the pleural wing process and the *subalare* in a similar position behind it. The basalare of the mesothorax (fig. 46 B, *Ba*) is an elongate sclerite hinged by its entire length on a wide lobe (*d*) of the upper margin of the pleuron, the posterior part of which is the wing process (*WP*) on which the second axillary (*2Ax*) is articulated. On the inner surface of the basalare is attached by a strong apodemal tendon a large flat muscle (G, 77) arising on the anterior part of the pleuron. The contraction of this muscle turns the basalar sclerite inward on its pleural hinge, and the movement is transmitted to the wing by the close connection of the basalare with the humeral angle of the wing base.

The subalare of the mesothorax is a small triangular sclerite (fig. 46 I, *Sa*), which in the bee is supported on a larger sclerite (*h*) that rests on the inflected upper margin of the epimeron (B, *Epm*). On the upper angle of the subalare is attached the tendon of a long muscle (I, 82) that arises ventrally on the mesothoracic coxa. The subalare is closely associated with the posterior basal process of the second axillary (B, *k*), so that the pull of its muscle is indirectly transmitted to the pivotal sclerite of the wing base. The function of the epipleural sclerite in the wing mechanism will be described later.

The Axillaries of the Hind Wing— In the simplified hind wing the humeral complex supports only the first two veins (fig. 47 A), and in the axillary region there is only a first, second, and third axillary.

The relatively small *first axillary* of the hind wing (fig. 47 A, B, *1Ax*) articulates on a detached marginal plate of the narrow metanotum (E, *d*), which probably represents the anterior notal wing process. Its posterior end rests in a depression (*f*) of the posterior angle of the metanotum, and its head abuts against the broad mesal end of the humeral complex (A). The irregular *second axillary* (A, B, *2Ax*) lies laterad of the first axillary and articulates by a ventral condyle in a socket on the dorsal margin of the metapleuron (C). The *third axillary* is a large, elongate sclerite (A, B, *3Ax*) lying transversely behind the second axillary between the edge of the metanotum and the fourth wing vein. As in the forewing the third axillary is the wing-flexing sclerite; its large three-branched muscle (B, 100) arises on the metapleuron and is attached on a small sclerite (A, *a*) in the wing membrane close to the inner end of the

123

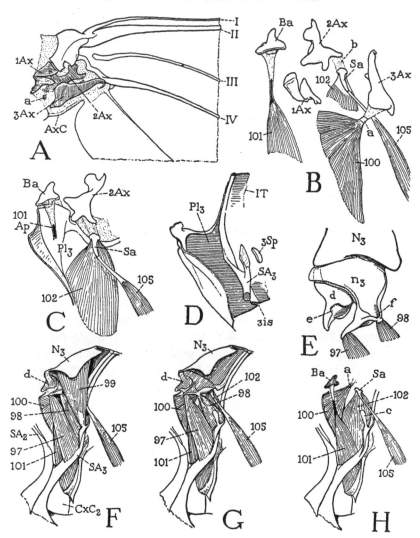

Fig. 47. Details of metathorax and base of hind wing of a drone.

A, base of right hind wing. B, axillary, basalar, and subalar sclerites. C, upper part of right pleuron, mesal, supporting basalare, second axillary, and subalare. D, upper part of pleuron supporting propodeal tergum. E, lateral part of metanotum, left side. F, lateral muscles in right side of metathorax. G, same with muscle *99* removed. H, same with muscles *97, 98, 99* removed, exposing muscles of basalare and subalare.

a, insertion point of muscle of third axillary in wing membrane; *b,* ligament connecting subalare with second axillary; *d,* free marginal plate of metanotum; *e,* lobe of *d* supporting first axillary; n_3, semidetached plate of metanotum. Muscles are explained in text. For explanation of abbreviations see page 133.

axillary, and hence pulls indirectly on the latter through the intervening membrane.

The Epipleurites of the Metathorax— In the metathorax the epipleural sclerites have the same relation to the hind wing that the corresponding sclerites of the mesothorax have to the front wing. The triangular basalare (fig. 47 B, *Ba*) sits on a marginal elevation of the metapleuron before the second axillary (C), and the small subalare (*Sa*) is attached to the concave pleural margin behind the second axillary. A single large basalar muscle (H, *101*) arises on the metapleuron behind the middle leg and tapers to a strong tendon attached mesally on the basalare. The inconspicuous subalare has two muscles, one a long fusiform muscle (B, C, *105*) arising ventrally on the hind coxa (fig. 37 A), the other a broad thick muscle (fig. 47 B, C, *102*) arising on the upper part of the metapleuron (H). The subalare is intimately associated with the second axillary of the wing base (B, *2Ax*) by a close membranous connection (*b*) with the posterior basal arm of the latter.

THE WING MECHANISM

In a study of the wing action we must distinguish between the movements of flight and the movements by which the wings are folded over the back when not in use or brought forward preparatory to flight.

The Wing Movements of Flight— The wing movements that produce flight include the up-and-down action of the wings and the sculling movements that give directed motion. With most insects the up-and-down wing strokes are produced by a vertical vibration of the back plates of the wing-bearing segments. As every entomologist knows, pushing a pin into the back of the thorax of a dead fly or bee at once raises the wings. This is because the downward pressure on the wing-bearing plate pushes down on the wing bases a very short distance mesad of the pleural wing fulcra. Conversely, a depression of the wings results from compression of the wing-bearing plate in the opposite direction, which restores the curvature of the back and pulls upward on the wing bases. In their vertical movements, therefore, the wings are levers of the first order, and the shortness of the power arms makes them extremely responsive to very small movements of the notal plates.

The wing-bearing plates of the back in most insects are sufficiently

125

thin and flexible to respond by a change of shape to the alternating pull of the vertical and longitudinal muscles attached on them. In the bee and related Hymenoptera, however, the back of the thorax is strongly convex, and the cuticle is so rigidly sclerotized that the wing-bearing plates cannot freely respond in the usual manner to the pull of the muscles. For this reason the notum of the mesothorax is cut by a transverse fissure (fig. 30, sf) through the posterior part of the scutum and is thus literally divided into a large anterior notal plate (fig. 49 C, $1N_2$) and a smaller posterior plate ($2N_2$). The dorsal part of the fissure serves as a hinge between the two notal plates, but the lateral parts widen into open membranous clefts (D, sf), which, by opening and closing, allow the otherwise rigid notum to respond to the alternating pull of the vertical and longitudinal muscles. The action is more easily demonstrated on a drone than on a worker bee.

A downward pressure on the mesonotum of the bee raises both pairs of wings if the wings on each side are connected but only the first pair if the wings are disconnected. Pressure on the metanotum alone, however, raises the hind wings. A longitudinal compression of the mesothorax depresses both pairs of wings if the wings are connected, but when the wings are unhooked no lengthwise compression of the thorax will depress the hind wings. From these facts it is evident that each pair of wings has its own elevator muscles, but that depression of both pairs of wings depends entirely on the longitudinal muscles of the mesothorax and on the connection of the hind wings with the front wings.

The cavity of the pterothorax, including that of the propodeum, is largely occupied by the huge dorsoventral and longitudinal muscles of the mesothorax (fig. 48). The thick lateral pillars of dorsoventral muscles (fig. 49 A, 72) have their upper attachments on the lateral parts of the scutum before the scutal fissure, and they go downward and posteriorly to their lower attachments on the ventral and lower lateral walls of the mesothorax. These muscles are the principal wing elevators of the mesothorax, but on each side of the segment is a smaller muscle (B, C, 75) arising on the episternum, the fibers of which converge to a short tendon attached on the margin of the posterior plate of the notum. This muscle probably acts as an accessory wing elevator by pulling directly down on the edge of the notum.

The longitudinal dorsal muscles of the mesothorax (fig. 48, *71*) are two great bundles of fibers attached anteriorly on the small prephragma of the mesothorax (*1Ph*) and on the strongly decurved surface of the scutum between the lateral tergosternal muscles (*72*). They extend posteriorly and downward through the metathorax to the posterior end of the propodeum, where they are at-

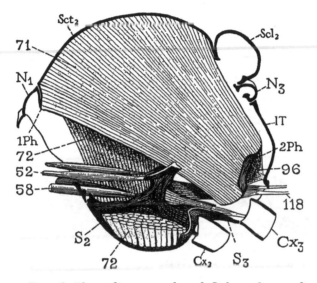

Fig. 48. The indirect muscles of flight and ventral thoracic muscles as seen in a median section of a drone's thorax.

52, 58, 118, ventral intersegmental muscles; *71,* indirect wing depressor muscle; *72,* indirect wing elevator muscle; *96,* retractor muscle of postphragma. For explanation of abbreviations see page 133.

tached on the mesothoracic postphragma (*2Ph*). Although there are no dorsal muscles in the metathorax, a pair of small fan-shaped muscles (*96*) from the end of the propodeum are attached on the posterior face of the phragma (fig. 49 A). It would appear that the propodeum has been added to the thorax in the clistogastrous Hymenoptera to give greater length and effectiveness to the mesothoracic wing depressors, which serve both the front and the hind wings.

The mesothoracic mechanism for the elevation and depression of the wings is illustrated diagrammatically at C and D of figure 49. At C the longitudinal muscles (*71*) are supposed to be in con-

traction: the scutal fissure (*sf*) is closed; the dorsum is at its maximal convexity; the notal margins are elevated to produce wing
depression. At D the vertical muscles (*72, 75*) are in contraction:
the anterior notal plate, being braced at *a* on the pleuron, is there-

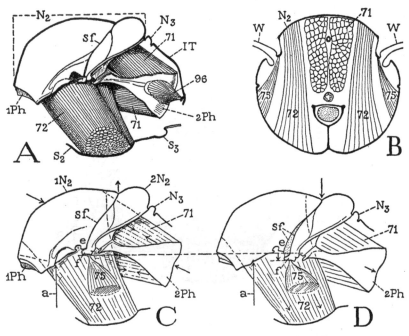

Fig. 49. Mechanism of wing movements by indirect muscles of flight.
A, indirect flight muscles in left side of pterothorax. B, diagrammatic cross
section of mesothorax showing position of muscles. C, diagram of dorsal
muscles (*71*) in contraction, wing articulations at *e* and *f* lifted, wings depressed.
D, same with vertical muscles (*72*) in contraction, notum depressed at arrow,
scutal fissure (*sf*) opened, wing articulations lowered, wings elevated.

71, indirect wing depressor muscle; *72*, indirect wing elevator muscle; *75*,
pleuronotal wing elevator muscle. For explanation of abbreviations see page
133.

fore pulled down posteriorly against the posterior plate; the scutal
cleft (*sf*) is opened and the adjoining margins of the notum are
depressed, whereby the wings are elevated. Experimentally the
same action can be produced by a downward pressure (arrow) on
the hinge line of the scutal fissure. The opening of the lateral clefts
of the fissure causes also a posterior displacement of the phragma,
and this movement of the phragma must stretch the wing-depressor

muscles (71) preparatory to their succeeding contraction. A length-wise compression of the thorax not only closes the lateral clefts of the mesoscutum, but causes the posterior margin of each cleft to slide under the anterior margin, thus giving the longitudinal muscles a maximal effectiveness for the downstroke of the wings.

The short metathorax contains no muscles corresponding with the dorsal longitudinals of the mesothorax and therefore contributes nothing to the downstroke of the wings. In each side of the meta-thorax, however, are three large vertical muscles arising ventrally on the metathoracic arm of the endosternum, which are attached dorsally on the notum (fig. 47 F, 97, 98, 99, G). These muscles evidently account for the independent elevation of the hind wings; a downward pressure at the region of their attachment on the metanotum at once turns the hind wings upward. The hind wings turn downward, however, only when they are hooked to the front wings and the mesothorax is compressed longitudinally.

The great size of the muscles that produce the up-and-down strokes of the wings shows that these movements are of major importance to the insect in flight. The driving power of the wings, however, is due to the twist of the wings on their long axes, by which the front margins go forward and turn downward accompanying the downstroke (fig. 50 F) and reverse during the upstroke (E). It is well known that the tip of a wing in motion while the insect is held stationary describes a figure-eight trajectory, just as does an oar used for sculling a boat from the stern. It is the epipleurites of the thorax and their muscles that produce these slight but important accessory movements of the wings.

On a dead bee it is easy to demonstrate that pressure against the basalar sclerite of the mesothorax turns the front wing forward and deflects its anterior margin. The basalare (fig. 50 D, Ba) is flexibly hinged on the upper edge of the episternum before the pleural wing fulcrum and is directly connected with the humeral angle of the wing base; its large muscle (77) is attached by a strong tendon on its inner surface (E). A contraction of the muscle, therefore, turns the basalare inward on the episternum (F) and deflects the anterior margin of the descending wing. Similarly, during the upstroke the subalare, which is connected with the pivotal second axillary sclerite, pulls downward on the posterior part of the wing base, so that the anterior wing margin turns upward and backward, relative to

129

the body, while the wing is ascending (E). The rotatory mechanism of the hind wing is the same as that of the forewing, the basalar and subalar sclerites of the metathorax (fig. 47 B, *Ba, Sa*) having the same relation to the wing base as in the mesothorax.

The insect mechanism of forward flight is thus not difficult to

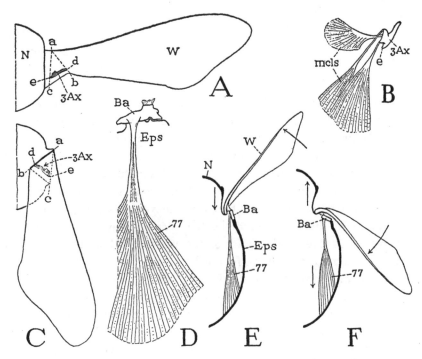

Fig. 50. Mechanism of wing flexion and of anterior deflection during flight. A, diagram of extended wing, axillary triangle (*abc*) flat. B, third axillary and its muscles. C, wing flexed posteriorly by dorsal revolution of third axillary (*3Ax*) on margin of notum (*e*), axillary triangle turned up and mesally. D, basalare and its muscles. E, wing elevated by depression of notum. F, wing in down stroke deflected anteriorly by pull of basalar muscle on basalare (*Ba*) accompanying elevation of notum.

understand, but many insects are capable of flying sidewise or backward. An expert at this kind of flying is the female *Syrphus* fly. Watch one hovering over a group of aphids intent on laying an egg among them. Suddenly she glides forward, then sidewise either right or left, and again she goes straight backward, all the time keeping her head toward the aphids. Even a honey bee collecting pollen, when she leaves a flower, darts quickly this way and that without any corresponding change in the position of her body. A study of the wing mechanism gives no evidence as to how these movements are produced. For maneuverability in flight the higher insects surpass any airplane yet invented, but they have had millions of years in which to perfect the details of their machinery.

The Wing Movements of Flexion and Extension— The more primitive of modern insects, such as the mayflies and dragonflies, have no mechanism for flexing the wings backward when at rest; at best they may bring them together over the body in a vertical plane. Most other insects, except the butterflies, fold the wings horizontally over the abdomen with the front wings on top, from which position the wings must be brought forward again preparatory to flight. The mechanism of flexion is simple and may be illustrated with a piece of stiff paper firmly held flat at one end and bent horizontally on itself; the bending necessarily produces a fold in the base of the paper. Conversely, if the paper is folded in this same way, it automatically bends backward at a right angle to its flat position. The flexing apparatus of the insect's wing acts on the second principle; by the *production* of a basal fold the wing is turned horizontally backward. The fold-producing mechanism is the third axillary sclerite and its muscles.

The action of the wing during flexion is readily demonstrated on most insects and may be represented diagrammatically as at A and C of figure 50. The wing (A, W) is hinged on the notum between a and c. A line of folding extends outward and backward from a to b, forming a basal triangle (abc) containing the axillary sclerites. The third axillary ($3Ax$) extends through the posterior part of the triangle; and if it is revolved upward and mesally on its articulation (e) with the notum, its distal end (d) will turn the triangle abc upside down as at C, with the result that the wing is automatically turned straight back on the fold ab. The third axillary is amply provided with muscles (B), but the action of the axillary is probably in most insects preceded by a relaxation of the basalar muscle, which allows the wing to turn a little backward and thereby give a sufficient preliminary tilt to the third axillary to allow its muscles to be effective. In the forewing of the bee this preliminary tilting of the third axillary is produced by a special mechanism involving the fourth axillary and the axillary lever.

131

The fourth axillary of the forewing (fig. 46 D, *4Ax*) is articulated on the margin of the mesonotum, and its distal end is pressed against the axillary cord (*AxC*). The head of the axillary lever (*axlv*) is exposed as a small sclerite (*s*) in the wing membrane immediately behind the fourth axillary; its internal part is articulated (F, *o*) on the base of the postphragma (*2Ph*), and its inner extremity gives attachment to a muscle (78) from the endosternum. A downward pull on the lever muscle depresses the lever on its fulcrum and causes the head to pull backward on the fourth axillary, and the axillary pushes back on the axillary cord. The wing gives a quick response by a slight backward movement, which brings the third axillary into a more nearly vertical position, making it thus more immediately responsive to the pull of its own muscles. The fourth axillary and its lever in the forewing of the bee, therefore, appear to be a starting apparatus for the flexor mechanism. The flexing of the forewing of the bee is difficult to observe because in the natural position of the extended wing (fig. 46 A) the third axillary (*3Ax*) lies beneath the large median plate (*m*). The action of the third axillary, however, can be demonstrated in the wing of a dead bee by turning this axillary upward on its notal hinge. The larger size of the drone makes the drone's wing a better subject for experimentation than that of the worker.

The flexing of the hind wing is produced entirely by the large third axillary (fig. 47 A, *3Ax*) and its muscle (B, *100*), since there is no fourth axillary or lever accessory. The axillary region of the hind wing, unlike that of the front wing, is approximately flat when the wing is extended (A). It is probable, therefore, that on relaxation of the extensor muscle the wing automatically turns backward sufficiently to raise the third axillary into a position in which its muscle can become at once effective.

The extension of the flexed wings preparatory to flight is in general produced by the basalar sclerites and their muscles. In a dead bee a slight pressure on the basalare of either wing, or a pull on its muscle, brings the wing out at a right angle to the body and deflects its anterior margin. In the mesothorax of the bee, however, there is a slender muscle from the pleuron attached directly on the humeral angle of the wing itself (fig. 46 C, *74*), and this muscle evidently must act as an accessory wing extensor. An extensor action of the subalar muscles cannot be demonstrated in the bee, though in some

132

other insects it would appear to extend the wing by flattening the fold in the base of the flexed wing.

The rate of the wing beat of the worker honey bee is given by Sotavalta (1947) as 208 to 277 per second, that of the drone as 220 to 233.

Explanation of Abbreviations on Figures 43–50

ANP, anterior notal wing process.

Ap, apodeme

1Ax, 2Ax, 3Ax, 4Ax, first, second, third, and fourth axillaries of wing base.

AxC, axillary cord.

axlv, axillary lever.

Ba, basalare.

C, costa, first wing vein.

Cu, cubitus, fifth wing vein.

Cx, coxa.

CxC_3, coxal cavity of hind leg.

Epm, epimeron.

epm, subdivision of epimeron.

Eps, episternum.

3is, third intersegmental groove of thorax.

M, media, fourth wing vein.

N, notum (a thoracic tergum).

Pcu, postcubitus, sixth wing vein.

1Ph, prephragma of mesothorax.

2Ph, postphragma of mesothorax.

Pl, pleuron.

pls, pleural sulcus.

PNP, posterior notal wing process.

R, radius, third wing vein.

SA_2, SA_3, mesothoracic and meta-thoracic arms of composite en-dosternum.

Sc, subcosta.

Scl, scutellum.

Sct, scutum.

sf, scutal fissure.

3Sp, propodeal spiracle.

IT, tergum of propodeum.

Tg, tegula.

V, vannal veins of wing.

vf, vannal fold of wing.

∗ VIII ∗

THE ABDOMEN

THE abdomen of an insect is a distinct anatomical section of the body projecting freely from the thorax, usually unsupported except by its attachment on the latter. Functionally it is an important body region because it contains the principal viscera, such as the stomach, the intestine, and the reproductive organs, and carries externally the organs of mating and egg laying. The base of the abdomen may be broadly joined to the thorax, though usually it is more or less narrowed. In the bee and other clistogastrous Hymenoptera, however, in which the first abdominal segment is incorporated into the thorax, the second segment is greatly constricted at its union with the propodeum, forming a slender *petiole* supporting the rest of the abdomen. In these insects, therefore, the functional abdomen is not entirely comparable to the abdomen of other insects. Structural alterations that bring about anatomical conditions in conflict with standardized nomenclature always create embarrassing problems for the descriptive anatomist. The abbreviated abdomen is sometimes called the *gaster,* but the most practical solution of the abdomen problem for hymenopterists is to use the term *abdomen* in a restricted sense for the third part of the insect body and to preserve morphological consistency by calling the first segment II and numbering the others in sequence. It must be remembered then that numerically each abdominal segment is one less than its number. In the bee the functional abdomen contains only nine segments, but the last is segment X.

Most adult insects have 10 segments in the abdomen, though in some orders there are eleven. The embryonic abdomen may show as many as 12 segmental divisions, and Nelson (1918) says that

134

there is evidence of 12 segments in the embryo of the honey bee but that the last two disappear before hatching. The terminal segment, called the *proctiger,* carries the posterior opening of the alimentary canal; the genital aperture of the female is usually in the ventral surface of segment VIII and that of the male in segment IX. Several of the posterior segments are commonly modified to serve the purposes of mating and egg laying, and these segments are frequently retracted into the segments before them, giving the abdomen the appearance of having fewer segments than it really possesses.

Most of the abdominal segments of adult insects except apterygotes lack appendages of any kind, but in the embryonic stage of some species each segment but the last bears a pair of small transient ventral protuberances that fall in line with the leg rudiments of the thorax, which fact suggests that the insects originally had legs along the entire length of the body. The adults of many insects have appendicular structures at the end of the body, which form an ovipositor in the female and copulatory organs in the male, and these structures are commonly regarded as derivatives of paired abdominal leg appendages of the eighth and ninth segments. There is still uncertainty, however, as to the true homologies of the external genital organs, the rudiments of which generally appear first in a late larval stage or on the pupa. Simple appendages on the tenth or eleventh segment, known as *cerci,* are absent in most Hymenoptera.

The individual segments of the abdomen are much simpler in structure than those of the thorax because they are not encumbered with either legs or wings, but if they once bore legs their simplicity is secondary. The sclerotized parts of each segment include a *tergum* above, and a *sternum* below. In some insects there are small lateral sclerites on each side between the terga and sterna, but it is doubtful if these sclerites in any case represent the pleura of the thorax; they are better termed *laterotergites* or *laterosternites.* The spiracles of the abdomen vary in position in different insects; in some they lie in the lateral parts of the terga, in others in the sterna or in lateral sclerites, or they may be in lateral or ventral membranous areas. The segmental rings of the abdomen are connected by infolded intersegmental membranes, which allow the abdomen to be contracted or extended lengthwise; lateral membranes connecting the terga and sterna allow dorsoventral expansion and compression. Movements in either direction or both directions may be utilized for respiration, though all insects do not make perceptible movements of breathing.

THE ABDOMEN OF THE BEE

The abdomen of the worker (fig. 51) is broad at the anterior end, which is abruptly narrowed at the petiole, and tapers to a point at the posterior end. The abdomen of the queen is similar in shape to that of the worker, though it is relatively larger, but that of the drone (fig. 52 B) is rounded at the posterior end. The six visible

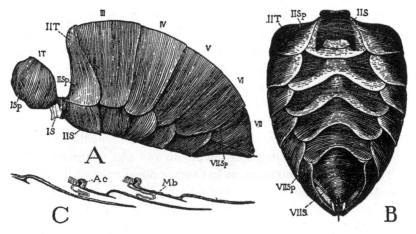

Fig. 51. The abdomen.

A, abdomen of worker, including the propodeal tergum. B, same, without propodeum, ventral. C, lengthwise section through three abdominal sterna, showing infolded intersegmental membranes.

For explanation of abbreviations see pages 166–167.

segments of the worker's abdomen are segments II to VII (fig. 51); the other three segments are concealed within segment VII and are much modified in connection with the sting. In the drone the exposed dorsal surface of the abdomen ends with the tergum of segment VIII (fig. 52 B), but the sternum of this segment (*VIIIS*) is concealed within that of segment VII, and the visible part of the venter terminates with sternum IX. The abdomen is thus divisible into a pregenital, or visceral, region of relatively unmodified segments and into a genital region in which the typical segmental structure is obscured in adaptation to the stinging mechanism in the female and the genital functions in the male.

136

The Pregenital Segments— The pregenital region of the abdomen has a fairly uniform segmental structure from segment III to segment VI in the female and segment VII in the male. The segment of the petiole and the segments of the genital region will need a special discussion.

The pregenital segments (fig. 51) have well-developed tergal and sternal plates, the terga overlapping the sterna along the sides of the abdomen. The spiracles (*Sp*) lie in the lateral parts of the terga. Anteriorly each tergum, except the first, is strengthened by an internal submarginal ridge, the *antecosta* (fig. 52 A, *Ac*), from which is given off on each side a short apodemal process (*a*) projecting forward into the preceding segment. The posterior tergal margin is extended as a free fold overlapping the front part of the following tergum, and the two terga are connected by an infolded intersegmental membrane. The sternal plates are smaller than the corresponding tergal plates, but they have a similar structure, except that the widely overlapping posterior edges of sterna III to VI in the worker (fig. 51 B) are deeply emarginate. The anterior margins of these sterna are thickened to form ridgelike antecostae, from which a pair of small apodemal processes project anteriorly (fig. 52 A, *b*). Each lateral margin supports a long lateral apodeme (*c*) that extends dorsally within the corresponding tergum. Both the tergal and the sternal apodemes give attachment to muscles that serve for respiration, as will later be shown. The ventral intersegmental membranes are reflected from the middle of each sternum to the anterior margin of the sternum following (fig. 51 C, *Mb*). There is thus a long, free posterior fold of each sternum that underlaps half the length of the following sternum. The connecting membranes themselves are folded, allowing extensive movements of the sterna on each other. The open spaces between the sterna are the wax pockets of the worker bee. The wax glands (fig. 52 A, *WxGld*) lie over these pockets on the anterior parts of the sterna of segments IV to VII.

The Petiole— The abdominal petiole is a highly important part of the body mechanism of the stinging Hymenoptera because it gives much freedom of movement to the abdomen on the thorax. Since the articulation of the abdomen on the thorax is between segments I and II of the primary abdomen, the principal muscles that move the functional abdomen are the intersegmental muscles between these two abdominal segments and not the usual thoracico-

137

abdominal muscles. Particular attention to the petiole as an important part of the organization of the bee was first given by Betts (1923), who fully described its structure and its principal muscles.

The narrow neck of the petiole segment is joined to the propodeum by a membranous conjunctiva (fig. 53 A, C), but the two

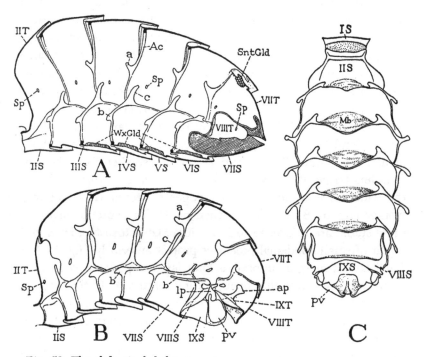

Fig. 52. The abdominal skeleton.
A, right half of abdominal wall of worker, internal. B, same of drone. C, abdominal sterna of drone, dorsal.

a, anterior tergal apodeme; *b*, anterior sternal apodeme; *c*, lateral sternal apodeme. For explanation of abbreviations see pages 166–167.

segments are hinged dorsally by a pair of articular points from the propodeum (C, *f*, *f*). Between the articular points on the dorsal surface of the petiole is a small domelike elevation of the membrane (*e*), which, when removed, is seen to cover a pocket from the cavity of the propodeum (F), the sclerotic floor of which on the petiole supports a median ridge (D, F, *r*). On the posterior end of this ridge are attached the tendons of two large muscles (*120*) that arise on the lateral walls of the propodeum (C). These muscles are the levators of the abdomen; their attachments on the posterior end of the ridge on the pedicel gives them a leverage on the propodeal fulcra (*f, f*). Opposed to the levators is a pair of long, ventral depressor muscles (E, *118*) arising anteriorly on the thoracic en-

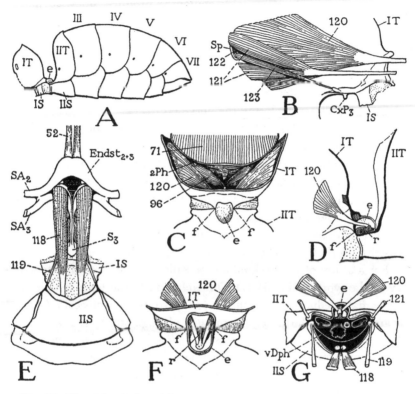

Fig. 53. The abdominal petiole.

A, abdomen of worker, lateral.   B, muscles in right side of propodeum, mesal. C, propodeum and petiole, dorsal wall of propodeum removed exposing muscles. D, levator mechanism of abdomen, left side of propodeum and petiole removed. E, ventral muscles of petiole, dorsal.   F, junction of propodeum and petiole, dorsal, upper wall of petiole pocket removed.   G, anterior end of petiole, showing muscle attachments.

e, membranous roof of petiole pocket; *f*, fulcra of movement of abdomen on propodeum; *r*, median ridge of petiole pocket. For explanation of abbreviations see pages 166–167.

dosternum (*Endst*) and attached posteriorly by tendons on the anterior margin of the petiole sternum (E, G, *IIS*). In addition to these dorsal and ventral muscles there are two lateral muscles on each side. One is a large two-branched muscle (B, *121*) arising on the side of the propodeum and attached by a strong tendon laterally on the end of the petiole tergum (G, *121*); the other (E, *119*), a more slender muscle, has its origin on the metathoracic sternum and is attached on the anterior lateral angle of the petiole sternum (E, G). These lateral muscles evidently produce sidewise movements of the abdomen.

The narrow cavity of the petiole gives passage to the aorta, tracheal tubes, the oesophagus, and the nerve cords from the thorax to the abdomen. It is divided by the ventral diaphragm (fig. 53 G, *VDph*), which begins in the metathorax, into a dorsal and a ventral channel, the latter enclosing the nerve cords.

The Pregenital Musculature— The musculature of a typical abdominal segment includes intertergal *dorsal muscles,* intersternal *ventral muscles,* and tergosternal *lateral muscles.* In addition to these major segmental muscles there are also muscles of the dorsal and ventral diaphragms and muscles of the spiracles. The segmental muscles follow the same plan in segments III to VI in both sexes of the bee (fig. 54 A, B), but the muscles of the drone (B) are much larger than those of either the worker (A) or the queen.

For a study of the abdominal musculature we may take segment IV of the worker (fig. 54 C). On each side of the tergum there is a large, flat median dorsal muscle (*144*) from the antecosta (*Ac*) of segment IV to the precostal margin of segment V and a smaller, more oblique lateral dorsal muscle (*145*) from the anterior part of the tergum to the antecosta of segment V. These muscles are tergal retractors. A third dorsal muscle (*146*), reversed in position, lies in the infold between the two terga and is a tergal protractor. It arises on the posterior part of tergum IV and goes forward to its attachment on the anterior apodeme of tergum V. The musculature of the sternum is similar to that of the tergum. Two long retractor muscles, one median (*152*), the other lateral (*153*), diverge posteriorly from the sternal antecosta of segment IV to the antecosta of sternum V. A short sternal protractor (*154*) lies in the intersternal fold, extending forward from the posterior part of sternum IV to the anterior sternal apodeme of segment V. There are likewise three tergosternal muscles on each side of the segment. Two of them cross each other obliquely in the female (*150, 151*) between the tergum and the sternum, but in the drone (B) they are parallel. These muscles serve to compress the abdomen dorsoventrally. The third lateral muscle (C, *149*) arises ventrally on the side of the tergum and extends

140

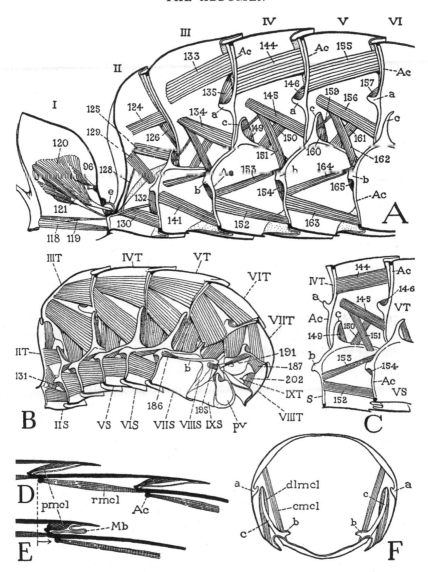

Fig. 54. The abdominal musculature.

A, muscles of right half of propodeum and first four abdominal segments of a worker, mesal. B, muscles of right half of abdomen of a drone, mesal. C, muscles of right half of segment IV of a worker, mesal. D, diagram of retracted abdominal terga. E, mechanism of protraction of segments. F, diagram of mechanism of dorsoventral compression and dilatation of a segment.

a, anterior tergal apodeme; *b,* anterior sternal apodeme; *c,* lateral sternal apodeme; *e,* membranous dorsal wall of petiole. For explanation of abbreviations see pages 166–167.

141

dorsally to its attachment on the upper end of the long lateral apodeme (*c*) of the sternum. This muscle serves to expand the abdomen dorsoventrally by separating the tergal and sternal plates.

The mechanisms of extension and dorsoventral expansion of the abdomen are shown diagrammatically at D, E, and F of figure 54. At D three consecutive tergal plates are in the usual position of retraction owing to the contraction of the retractor muscles (*rmcl*). At E the second tergum has been protracted posteriorly by the contraction of the reversed protractor muscle (*pmcl*), throwing the intertergal membrane (*Mb*) into a wide fold. The dorsoventral expansion of the abdomen is effected in the same way by the reversed tergosternal lateral dilator muscles (F, *dlmcl*) attached on the upper ends of the long lateral apodemes (*c*) of the sternum. The contraction of these muscles pushes the sternum downward as far as the membranous tergosternal conjunctiva will allow; the direct lateral compressor muscles (F, *cmcl*) reverse the movement. The respiratory movements of the bee are served by these mechanisms of expansion and contraction, but the movements of the segmental plates on each other probably have other purposes, and the strong musculature of the abdomen of the male (B) suggests that such movements are of particular importance to the drone.

The Genital Segments— The genital region of the abdomen includes segments VII, VIII, and IX, all of which are more or less modified as compared with the pregenital segments, and they are quite different in the male and the female.

In the drone segment VII has essentially the same structure and tergal musculature as the preceding segments (fig. 54 B), but because of the compression of the sterna of the segments in front of it, its own sternum (*VIIS*) lies anterior to the tergum (*VIIT*). The dorsum of the exposed part of the male abdomen ends with the strongly decurved tergum of segment VIII (*VIIIT*), but the reduced sternum of this segment (*VIIIS*) is normally concealed within that of segment VII, so that the externally visible part of the abdominal venter terminates with sternum IX, which is a well-developed crescent-shaped plate (fig. 58 I, *IXS*). The tergum of segment IX, on the other hand, is represented only by a pair of small sclerites (*IXT*) concealed within tergum VIII in the membrane at the sides of the anus-bearing proctiger (*Ptgr*). Each of these small sclerites, however, supports internally a large apodeme (fig. 54 B,

IXT) giving attachment to the dorsal muscles of its segment (*187, 202*). At each side of the ninth sternum is a small plate (fig. 58 I, *lp*), and between the two is a pair of larger free lobes (*pv*). These structures are the only external parts of the male genital organ; between the median lobes is a large aperture (*Phtr*), the *phallotreme*, from which the complex intromittent organ is everted at the time of mating. The tenth abdominal segment is represented only by the short, tubular anus-bearing proctiger (fig. 58 I, *Ptgr*) projecting from beneath the eighth tergum.

In the worker and the queen the exposed part of the abdomen ends with the conical seventh segment (fig. 51 A, *VII*), from the apex of which the point of the sting may be seen projecting. The tergal and sternal plates of segment VII are not essentially different from those of the preceding segment, except that the free posterior part of the sternum, instead of being deeply emarginate, is produced into a wide triangular extension.

Segments VIII, IX, and X and the sting are all withdrawn into segment VII (fig. 59), where they are contained in a large chamber, the walls of which are the intersegmental membranes inflected forward from the ends of the tergal and sternal plates of the seventh segment to the base of the retracted eighth segment. Segment VIII is a complete annulus, but its walls are membranous except for two lateral plates (*Lsp*) that contain the last pair of spiracles (*Sp*), and which therefore are sclerotic remnants of the eighth tergum (*VIIIT*). These spiracle-bearing plates are so closely associated with the sting that in apicultural terminology they are commonly known as the *spiracle plates* of the sting. On the dorsal part of each plate is attached a large intertergal muscle (fig. 55 B, *178*) from tergum VII, and on the ventral angle an intersegmental sternotergal muscle (*184*) from sternum VII. Muscles to parts of the sting derived from segment VIII arise on the mesal surface of the spiracle plate (*187, 188*), and on the ventral margin arises a long muscle (*192*) that goes to the triangular plate of the sting (fig. 59, *Tri*).

The ninth segment furnishes the principal support for the sting, but it is hardly to be recognized as a segment, since its only sclerotized parts are two lateral tergal plates (fig. 59, *Qd*) that form an essential part of the sting apparatus and are termed the *quadrate plates* of the sting. Below the quadrate plates is a pair of *oblong plates* (*Ob*), which also belong to segment IX; their lower edges

143

are united by the membranous venter of the segment, which is
arched over the base of the shaft of the sting (fig. 60, *IX.V*). The
muscles of the female genital segments will be described later in
connection with the sting and its mechanism. The anus-bearing
tenth segment of the female (fig. 59, *Ptgr*) has the same form as

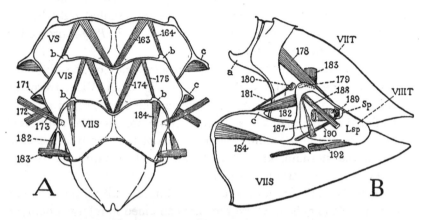

Fig. 55. Terminal abdominal segments of a worker, with muscles.
A, ventral musculature of segments V, VI, VII, dorsal. B, muscles of right
half of segment VII and of spiracle plate of segment VIII, mesal.
For explanation of abbreviations see pages 166–167.

that of the male, and in the retracted position lies in the sting
chamber beneath the seventh tergum. In a depression of the inner
wall of the chamber below the base of the sting is the opening of
the female reproductive system.

THE WAX GLANDS AND WAX MANIPULATION

The anterior parts of the sterna of segments IV to VII of the abdomen that form the true ventral walls of their segments, but are
concealed above the free underlapping posterior parts of the preceding sterna, are each characterized in the worker by the presence
of two large, oval, polished surfaces framed in dark marginal bands
and separated by a broader median space (fig. 56 A, *Mir*). These
polished areas are known as the *wax plates,* or *wax mirrors;* above
them are situated the four pairs of wax glands of the honey bee
(fig. 52 A, *WxGld*), which are merely specialized areas of the
epidermis. Lying over each gland is a large cellular mass composed
of fat cells and oenocytes (fig. 56 B, *FtCls, Oen*).

144

From the work of Dreyling (1903) and later writers it is well known that the thickness of the gland cells varies with the age of the bee or according to the time at which the bee is actively producing wax, as shown at C of figure 56 from Rösch (1927a). In young, freshly emerged workers the epidermis over the wax mirrors is an ordinary flat epithelium, but as the activities of the bee increase, the

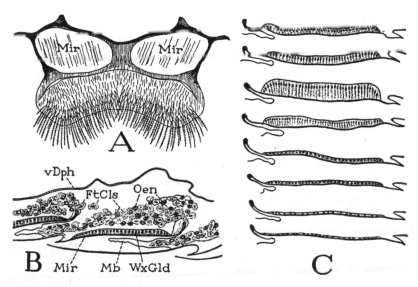

Fig. 56. The wax glands.

A, sternum of segment VI of worker, ventral, showing polished "mirrors" beneath wax glands. B, lengthwise section through two wax glands with overlying masses of fat cells and oenocytes (from Rösch, 1930). C, stages in the development and regression of a wax gland (from Rösch, 1927a).

For explanation of abbreviations see pages 166–167.

cells elongate vertically and clear spaces appear between them. At the height of wax secretion the gland cells are greatly lengthened. In old age again most of the cells shrink, and in overwintering workers the epidermis on the wax plates becomes a simple sheet of nucleated plasma in which even the cell boundaries are lost. According to Reiman (1952), the actively secreting gland cells have a longitudinally fibrillated appearance and contain numerous basophile granules. They are richly supplied with tracheae from which proceed tracheoles that penetrate into the cells and into the intercellular spaces.

The cuticle of the wax plates is described by Reimann (1952) as having a stratified fibrous structure in which the fiber layers cross one another in different directions. Treatment with clearing reagents brings about a change in the appearance of the texture; the fibers that in untreated preparations are closely adherent now separate into a latticework (*Gittersystem*) with wide interspaces. Beneath the exocuticle Reimann describes a strongly staining layer that he calls the mirror layer (*Spiegelschicht*). It is first seen in late pupal stages shortly before emergence and attains its maximal development during wax secretion; in old bees with degenerating or degenerated glands it is absent. This layer, Reimann says, is evidently not the endocuticle; structurally it consists of a porous meshwork, and it plays an important part in the excretion of wax. Immediately under the mirror layer is a zone of vertical fibers, the *Filterstäbchen* of Reimann, that takes stain strongly during wax secretion. Seen from the inside the fibers appear as large numbers of black points within the limits of the gland cells, but it is not certain whether or not they are continuous with the plasmatic fibers of the cells.

The masses of fat cells and oenocytes lying over the wax glands evidently must have some functional relation to the gland cells. Reimann (1952) follows in detail the apparent correlation in the activities of the oenocytes and the wax-gland cells. He finds that the oenocytes increase and attain their maximal size when the gland cells have reached their greatest height during wax production. The histological picture, he says, suggests also that the oenocytes grow at the expense of the fat cells. Evidence was earlier given by Rösch (1930) that in older bees both fat cells and oenocytes are bodily absorbed by the cells of the wax glands. The fat cells in this case, he points out, do not discharge their products into the blood, but give them by direct contact to the growing wax glands, and the dark granular contents of the oenocytes, he says, can be seen for a long time in the gland cells.

In their study of the dermal wax glands of the cockroach, Kramer and Wigglesworth (1950) suggest that the oenocytes, which in the cockroach also are associated with integumental wax glands, are the most probable source of the wax discharged as a film on the surface of the cuticle. Though there is no definite proof that excreted wax is first elaborated in the oenocytes, the idea that the oenocytes are the source of wax is in harmony with the contention

146

of Hollande (1914) that the oenocytes of insects in general are organs for the metabolism and conservation of wax.

Finally we have the problem of how the wax gets through the cuticular wax plates. The plates have been described as being perforated by "pore canals"; Lewke (1950) says the pores are clearly to be seen in tangential sections of the cuticle, and Dreyling (1903) estimated that there are 30 to 50 of them corresponding to each cell of the gland. On the other hand, Reimann (1952) states positively that no pores are visible in the wax plates of the honey bee. He postulates that the wax components are transported through the gland cells in a protein medium, which is broken down and liberates the wax molecules before entrance into the *Spiegelschicht*. The wax is then carried through the cuticle in solution in some special secretion in the form of molecules or molecule chains, and finally on the outer surface of the cuticle the molecules condense and the wax hardens. At least we do know that somehow the wax gets through an apparently nonporous cuticle and in the intersternal pockets of the abdomen forms the familiar wax scales with which the bees construct their comb.

The method by which the bee removes the wax scales from the abdominal pockets and the manipulation of the wax in comb building have been described in detail by Casteel (1912a). Formerly it was thought that the pollen presses on the hind legs were instruments for the removal of wax, and these structures were then known as the "wax shears." Casteel, however, found that it is the brushes on the basitarsi of the hind legs that the bee uses for extracting the scales, and his observations were confirmed by Rösch (1927a). The hind legs, one at a time, are pushed back with the tarsal brush against the undersurface of the abdomen, and the spines on the end of the brush disengage a scale from its pocket; the scale usually adheres to the brush. One scale, Casteel observes, may be taken from one side and the next from the other side, or two or three may be removed in succession from the same side. The leg bearing a scale is then bent forward, and the wax is taken off by the mandibles, perhaps with the assistance of the forelegs, and is masticated in the jaws. Usually the bee that has produced the wax now proceeds to deposit it on the comb under construction.

147

THE SCENT GLAND

When a worker bee extends the tip of her abdomen, particularly when she bends the apical segment downward, there is exposed on the base of the last tergal plate a wide pale band, which ordinarily is concealed beneath the overlapping part of the tergum in front. This pale area has been described as the "intersegmental membrane," but actually, as shown by Jacobs (1924), it is a weakly sclerotized basal part of tergum VII (fig. 57 A) to which the true intersegmental membrane (Mb) is attached. Beneath it lies the abdominal scent gland.

A dorsal view of tergum VII of the worker (fig. 57 A) shows that the basal pale area is margined anteriorly by the line of the antecosta (Ac), which becomes very weak medially, and is separated posteriorly from the sclerotized, hairy part of the tergum by a sharp transverse line (c). Between this line and the antecostal line are two other transverse lines (a, b). The meaning of the surface lines is seen in a longitudinal section of the tergum (C). The antecosta (Ac) is a very slight ridge on the anterior tergal margin, while the line a behind it is formed by a much larger submarginal internal ridge. The line b is the crest of a transverse fold, which forms the posterior wall of a deep canal (Can) across the base of the tergum. Behind the canal the elevated tergal surface slopes gently back to the line c, where a groove separates the pale area of the tergum from the dark, hairy exposed area. The scent gland (SntGld) lies below the anterior half of the pale area, and its ductules open into the basal canal (Can). From the precostal lip of the canal is reflected the intersegmental membrane, which is doubled in a large fold (Mb) between terga VI and VII, which allows tergum VII to be sufficiently protracted to expose the scent-gland area on its base.

Tergum VII of the queen (fig. 57 B) is larger than that of the worker and has none of the special features of the latter. Its overlapped basal part is paler than the exposed part and is devoid of hairs; anteriorly it is margined by a strong internal antecosta (Ac).

The scent gland of the worker is a mass of large glandular cells lying beneath the canal of the tergum and the anterior half of the elevated area behind it (fig. 57 C, SntGld). The structure of the gland and its cells has been well described by McIndoo (1914a) and by Jacobs (1924). Each cell of the gland has a delicate, in-

148

dividual, hairlike duct (E, *Dct*), which arises in a clear space, or ampulla (*Amp*), within the cell and opens through the cuticle on the floor of the canal (D, *Dcts*). In a dissection of a worker, when the gland cells are cleared away, the ducts appear as a great number

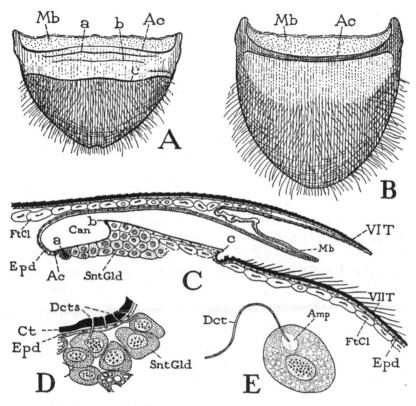

Fig. 57. The scent gland.

A, abdominal tergum VII of a worker. B, same of a queen. C, lengthwise section through base of tergum VII of worker, covered by tergum VI. D, group of scent-gland cells and ducts (from McIndoo, 1914a). E, a single gland cell and its duct (from McIndoo, 1914a).

a, line of strong submarginal inner ridge on base of tergum VII of worker; *b, c,* margins of elevation on tergum VII of worker. For explanation of abbreviations see pages 166–167.

of minute filaments attached to the inner surface of the cuticle. McIndoo says there are from 500 to 600 of them. The ducts are said by Jacobs to dissolve completely in 20 per cent caustic, and therefore to be nonchitinous. McIndoo reported just the opposite results from treatment with saturated caustic potash for a few hours and concluded that the ducts are chitinous. In general, ducts of unicellular dermal glands of insects are lined with a delicate cuticular intima continuous with the external cuticle (see Richards, 1951, p. 213).

The scent gland is absent in the drone of the honey bee. According to McIndoo, it is present in the queen, but Jacobs says the queen has no scent gland; its absence in the queen would seem to be indicated by the simple structure of tergum VII. A corresponding scent gland, according to Jacobs, is present also in *Apis dorsata, florea, indica, adamsoni,* and *unicolor* and shows a progressive development in this order from *dorsata* to *mellifera.*

There appears to be no doubt that this gland of tergum VII in the worker, known as the gland of Nassanov, is the chief source of odor emitted by honey bees. If the secretion is discharged into the depression on the base of the tergum, it becomes exposed when the tergum is protracted and depressed and can then be dispersed by air currents blown backward from the fanning wings. Kalmus and Ribbands (1952) suggest that there may be other sources of bee odor. They furthermore show that bees of each hive have a distinctive odor, by which they distinguish their companions from bees of another hive. Nixon and Ribbands (1952) find that this colony odor is not inherited but is acquired by the bees of each hive. They offer in explanation of its specific quality the fact that all the bees of a colony, by mutual exchange of food, receive the same food material, which gives rise to similar waste products and thus creates a distinctive hive odor. (See also Ribbands, 1955.)

OTHER EPIDERMAL GLANDS

In many insects unicellular glands of the epidermis are variously distributed over the body. In the cockroach such glands are said to produce a waxy secretion that forms a film on the outside of the cuticle. According to Heselhaus (1922) and Jacobs (1924), unicellular dermal glands occur among both social and solitary Apidae on all parts of the body, but particularly on the abdomen. Heselhaus reports that about 20 such glands are present in the queen bee on the last two pairs of wax plates, 10 on the last exposed tergum near the spiracles, about 50 on the preceding tergum, and 10 to 15 on the next tergum in front, and that scattered pores occur on the other terga. Jacobs says all social and solitary Apidae examined have on the abdomen a more or less strongly developed system of ampullar gland cells very differently distributed in different genera. They occur principally, however, on the lateral parts of the anterior borders of the segments, where they appear to be oil glands (*Schmierdrüsen*), and are most abundant on the last three segments.

It is well known that worker bees habitually lick the queen with their tongue, and it seems to be well attested by the work of Butler (1954a, 1955) that the workers thus obtain some substance from the queen by which they keep informed of the presence of the queen in the hive. If they are prevented from access to the queen, even though allowed to "see" her, they at once behave as if the colony were queenless and begin building new queen cells. Butler observes that after a worker has licked the queen she offers food to other workers, and he suggests that it is by means of food exchange that the "queen substance" is distributed through the colony and that by this means all the workers are kept informed of the presence of the queen. If the bees do obtain some such substance from the queen, it must be an exudation from dermal glands, and the quantity must be very small. It is not clear, therefore, how the substance is transferred from the tip of the tongue to the base of the proboscis where the proffered food is regurgitated. However, we may suppose that the exudation licked from the queen is dissolved in saliva discharged from the tongue canal, which is then sucked up to the mouth through the food canal. The aperture of the salivary canal on the base of the labellum (fig. 26 G) is too closely guarded by hairs to serve as an intake orifice. In any case the worker bees must have an extremely delicate sense of "taste" if the lickings from one queen can control the entire colony, as Butler's experiments seem to demonstrate.

THE MALE GENITALIA

The external genital organs of most adult male insects are highly complex structures including not only a specific organ of intromission but variously developed clasping accessories. In their development, however, as has now been shown in most of the major insect orders, the male organs appear first as a single pair of small protuberances on the venter of the ninth or tenth abdominal segment, between the bases of which a tubular ingrowth of the ectoderm becomes the common genital exit duct known as the *ductus ejaculatorius*. As these primary genital, or *phallic*, rudiments lengthen,

151

each splits terminally into two branches; the outer branches are appropriately called *parameres*, and the median branches may be termed *mesomeres*. In such insects as cockroaches four phallic parts, or *phallomeres*, simply surround the opening of the ejaculatory duct, though they become variously developed. In most of the insects above Orthoptera the parameres become claspers, while the mesomeres unite with each other by their dorsal and ventral edges to form a tubular organ known as the *aedeagus*. The external wall of the aedeagus becomes sclerotized, but the inner wall usually remains soft, forming a membranous inner tube, the lumen of which is continuous with that of the ejaculatory duct opening through the primary *gonopore* into its inner end. The inner tube of the aedeagus thus becomes the exit conduit for the spermatozoa, and usually during mating it is everted, bringing the gonopore to its tip, and so becomes the functional intromittent organ, or *penis*, of the male insect. We may, therefore, classify the phallic structures into *ectophallic* parts, which include the parameres and the aedeagus, and into an *endophallus*, which is the eversible inner tube of the aedeagus. The parameres may become entirely separated from the aedeagus, or the three parts may remain united on a common *phallobase* as they do in most Hymenoptera. In the honey bee, however, the ectophallic parts are greatly reduced, and the endophallus is elaborated into a large and highly complex eversible intromittent organ.

The development of the external genital structures in the drone bee has been described in detail by Michaëlis (1900) and by Zander (1900). The principal stages in the growth of the organs are readily seen in whole specimens. The earliest stage appears in a late larval instar as a small oval disc with two slight thickenings on the ventral surface of the ninth abdominal segment (fig. 58 A, *gd*). The next stage is to be seen in the succeeding instar developing within the cuticle of the other. Beneath the outer disc is an oval pit (B) from the anterior wall of which projects a pair of small rounded lobes (*phL*), which are the *primary phallic rudiments*, represented by the pair of slight thickenings in the outer disc. With further development the primary lobes unite at their bases and divide distally into four secondary lobes (C). These secondary lobes, therefore, may be identified as the paramere (*Pmr*) and mesomere (*Mmr*) branches of the phallic rudiments in other insects. At the following transformation to the pupa, the immature phallus is exserted from the containing pit and now appears (E) as a thick body with a broad base (*PhB*) supporting the four apical lobes; it may be seen projecting from the end of the pupal abdomen (D, G) between the proctiger (*Ptgr*) above and the ninth sternum (*IXS*) below.

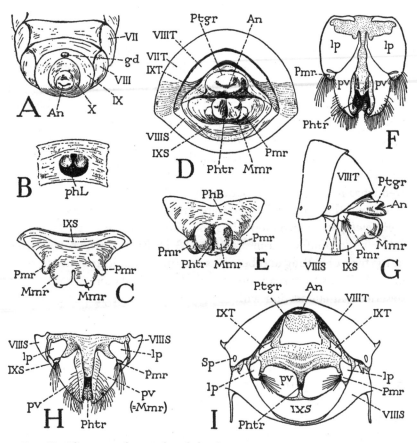

Fig. 58. The external genitalia of the drone.

A, end of abdomen of male larva. B, primary phallic lobes in pit of epidermis beneath larval cuticle. C, phallic lobes of pupa, ventral. D, end of pupal abdomen, posterior. E, same as C, dorsal. F, phallic organ of adult *Apis florea* F. G, end of abdomen of male propupa of honey bee. H, phallic organs of adult drone, dorsal. I, terminal abdominal segments of drone, posterior.

For explanation of abbreviations see pages 166–167.

At this stage in most insects the mesomere lobes have united to form an aedeagus; in the bee they remain separate, but between them is a slitlike aperture, the phallotreme (fig. 58 D, E, *Phtr*), which is the external opening of a deep endophallic pouch. The pupal organ shows no change externally during the pupal period, but within the pupal cuticle the imaginal organ (G) completes its development and attains the final structure before the pupal cuticle is shed.

In the adult the ectophallic parts have assumed definite shapes (fig. 58, H, I). The mesomere lobes are relatively large and are now commonly termed the *penis valves* (*pv*) because they close over the phallotreme (*Phtr*), from which the endophallic penis is eversible. The parameres (*Pmr*) are small lobes bearing brushes of long hairs borne on a pair of parameral plates (*lp*) in the broad base of the phallus. A structure more typical of the phallus of other Hymenoptera is seen in *Apis florea* (F), in which the parameral plates (*lp*) occupy a large part of the phallobase. The parameres of *florea* resemble those of *mellifera,* but in some Hymenoptera they are large free lobes movable on the parameral plates.

Though the ectophallic parts are greatly reduced in the drone of *Apis mellifera,* the endophallus, on the other hand, is developed into a huge complex structure (fig. 103 D, *Enph*), which is everted at the time of mating and becomes the functional intromittent organ, or penis. The structure of the endophallus will be more appropriately described in connection with the internal male organs of reproduction.

THE STING

The stinging organ of the bees and the wasps is an elaborated ovipositor, and its parts are readily identified with those of the ovipositor of such insects as the Orthoptera, the Hemiptera, and the nonstinging Hymenoptera. Most of these insects use their ovipositor for thrusting their eggs into crevices, under bark, or into the ground or for making slits or holes in the branches of trees in which to insert the eggs, while parasitic Hymenoptera use their ovipositors for puncturing the bodies or eggs of other insects, in which they deposit their eggs. The bees and the wasps inherited an ovipositor from their nonstinging ancestors, but when they took to rearing the brood in combs, they ceased to need an organ for egg-laying purposes; the ovipositor then served efficiently as an instrument of defense and, when supplied with venom, became a sting. It is very probable that the glands associated with the bee's sting are the usual

154

accessory genital glands of other female insects, which ordinarily discharge, into the base of the ovipositor, a secretion used for cementing the eggs to a support or for forming egg cases.

The sting of the honey bee is contained within the large sting chamber in the end of the abdomen enclosed between the tergal and sternal plates of segment VII (fig. 59). The sting chamber, as

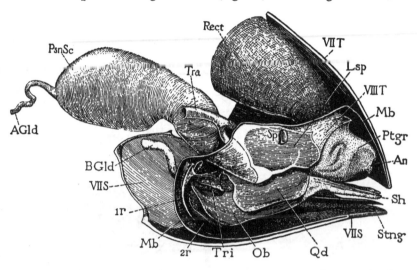

Fig. 59. The sting of a worker and associated structures in the sting chamber, exposed by removal of left wall of segment VII.

For explanation of abbreviations see pages 166–167.

already noted, contains also the reduced and modified sclerites of segments VIII and IX and the apical anus-bearing lobe, or proctiger, which represents the tenth segment; in fact, the sting chamber is formed by a retraction of these segments into segment VII.

The Structure of the Sting— The complex stinging organ of the bee is an intricate mechanism, and we cannot understand its action without knowing every detail of its structure. The entire organ (figs. 60, 61 A) includes two sets of parts that are anatomically and functionally distinct. One is the large basal part, which is the principal *motor apparatus*, made up of several pairs of plates by which the sting is attached within the sting chamber of the abdomen; the other is the long tapering shaft, which is the *piercing instrument* and is alone protracted from the abdomen during the act of stinging. The two parts are connected by a pair of curved basal arms, each of which is composed of two closely adherent rami.

155

The basal part, or motor apparatus, of the sting presents on each side three plates (figs. 60, 61 A). Uppermost is a large plate (Qd), which in bee literature is called the *quadrate plate*, but which, as we have seen, is a lateral remnant of the tergum of segment IX. The dorsal margin of the quadrate plate is produced into the body cavity as a flat apodeme (Ap), which gives attachment to muscles.

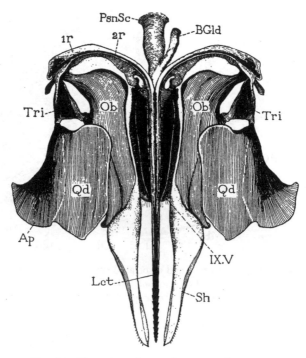

Fig. 60. The sting of a worker, ventral.
For explanation of abbreviations see pages 166–167.

Below the quadrate plate, and partly overlapped by its ventral margin, is a horizontally elongate plate (Ob) known as the *oblong plate*. In front of the quadrate plate and above the anterior end of the oblong plate is a small *triangular plate* (Tri), which is articulated by the dorsal angle of its base (fig. 61 A, c) with the quadrate plate and by its lower angle (d) with the upper edge of the oblong plate. The apex of the triangular plate (b) is continuous with the first ramus ($1r$) of the arc that connects the shaft with the motor apparatus, and the anterior end of the oblong plate is continuous with the second ramus ($2r$). The lower edges of the two oblong plates are connected by a convex setose membrane, which is the unsclerotized ventral wall of segment IX; posteriorly it ends in a free hairy lobe (*IX.V*) over the base of the sting. Projecting from the distal ends of the oblong plates are two long, soft tapering appendages (*Sh*), which normally clasp the shaft of the retracted sting and thus form a *sheath* for the latter.

The shaft of the sting consists of three closely appressed parts that together taper to a sharp point (fig. 61 A). The unpaired upper component is the *stylet* (*Stl*); the paired lower elements are the *lancets* (*Lct*). Proximally the stylet is expanded in a bulblike enlargement (*blb*), the cavity of which is open below, and is continued as a ventral groove to the end of the stylet (H, *Stl*). The bulb is covered dorsally by the membranous venter of segment IX that unites the lower edges of the oblong plates (A, *IX.V*). The lancets are slender rods armed distally on their lateral surfaces with barblike teeth pointed anteriorly (B); proximally they are directly continuous with the first rami (A, *1r*) of the basal arms of the sting, which are connected with the triangular plates of the motor apparatus. The second rami (*2r*) from the anterior ends of the oblong plates are attached to the basal angles of the bulb (D, *blb*). The sac of the poison glands (A, *PsnSc*) opens by a narrow neck into the base of the bulb cavity. Above the base of the bulb is a small, forked, apodemal sclerite, the *furcula* (C, F, *Frc*, G), which gives attachment to important muscles of the sting (C, *197*).

The slender lancets are held close against the undersurface of the stylet by grooves that fit over tracklike ridges of the latter (fig. 61 H), and the same interlocking device is continued upon the basal rami, the first ramus being grooved to its upper end (E, *j*), the second having a strong ridge along its convex margin (D, *i*). The lancets are thus able to slide freely back and forth on the underside of the stylet. Between the stylet and the lancets is the poison canal of the sting (H, *pc*), which expands proximally into the cavity of the bulb, and here receives the liquid poison from the great poison sac opening into the base of the bulb. Each lancet bears proximally on its upper edge a pouchlike valvular lobe with its concavity directed backward (E, *Vlv*); the two lobes project into the lumen of the bulb, where they serve to propel the poison through the poison canal when the sting is in action.

As the sting lies in the sting chamber of the abdomen (fig. 59),

157

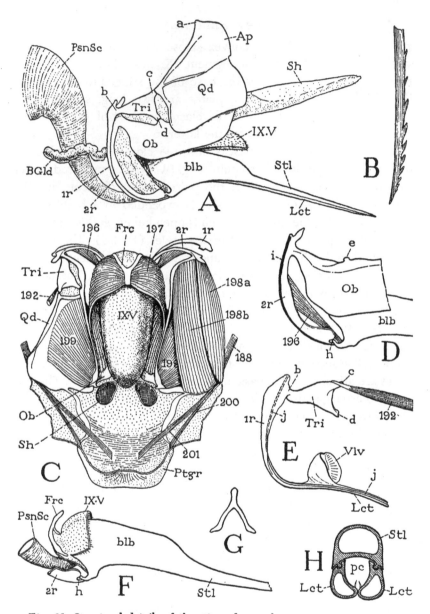

Fig. 61. Structural details of the sting of a worker.

A, outline of left side of sting. B, barbed end of a lancet. C, muscles of sting, dorsal. D, base of oblong plate and bulb connected by second ramus. E, triangular plate and base of lancet connected by first ramus. F, bulb of stylet articulated on second ramus with the associated furcula. G, furcula. H, cross section of distal part of sting shaft.

Alphabetical lettering and muscles are explained in text. For explanation of abbreviations see pages 166–167.

the quadrate plates are partly overlapped by the spiracle plates (*Lsp*) of segment VIII. The quadrate plate and the spiracle plate of each side are connected by an infolded intersegmental membrane, but the apodeme of the quadrate plate (fig. 61 A, *Ap*) is articulated by its dorsal angle (*a*) to the posterior end of the overlapping spiracle plate (fig. 62 A).

The Glands of the Sting— Two glands are associated with the base of the sting. The principal gland consists of a pair of long, slender, much-convoluted tubules lying in the posterior part of the abdomen (fig. 104, *AGld*). Each tubule ends with a small glandular enlargement, and the two tubules unite in a short common duct that opens into the anterior end of the poison sac (*PsnSc*), which by its tapering posterior end discharges into the cavity of the bulb of the sting shaft. This gland is commonly known as the *acid gland* of the sting; the secretion is said to consist principally of formic acid. The epithelial walls of the poison sac have a thick, laminated cuticular intima thrown into numerous high folds. In the neck of the sac the folds form regular transverse rings, holding the neck rigidly open. The sac walls have no muscles, and the poison liquid therefore is not expelled by contraction of the sac; it is driven through the canal of the sting by the action of the lancets and their valves. The secretion of the acid gland is the venom of the bee's sting.

The second gland of the sting is a short, thick, slightly convoluted, opaquely whitish tube (figs. 61 A, 104, *BGld*), known as the *alkaline gland* because of the nature of its secretion. Its walls consist of a thick epithelium of distinct cells lined by a thin cuticular intima. This gland opens ventrally at the base of the sting.

The function of the alkaline gland has been a subject of much discussion and difference of opinion ever since the gland was first described by Dufour in 1841; a good historical review of the controversy is given by Trojan (1930). It once seemed established by the experiments of Carlet (1890) on the toxic effects of the sting venom on flies that the poison acts with full strength only when the secretion of the alkaline gland is injected with that of the acid gland. Trojan (1930), however, has clearly shown that the alkaline gland does not open into the lumen of the sting; it extends ventrally past the base of the bulb and discharges into the sting chamber of the abdomen below the sting at the point where the lancets begin

159

to diverge forward. The secretion of this gland, therefore, could not be mixed with that of the acid gland and must have some specific function of its own. Other writers have suggested that the alkaline secretion may serve as a lubricant for the moving parts of the sting or neutralizes the remains of the acid secretion in the sting after stinging. The alkaline gland is present in other Hymenoptera, and in the solitary bees it is relatively of much greater size than in *Apis*. Trojan concludes that the alkaline gland simply retains its primary status as an accessory genital gland and suggests that its secretion in the queen may be discharged as a protective covering on the eggs issuing from the vagina, or perhaps serves as an adhesive for attaching the deposited eggs to the wall of the comb cell.

In addition to the two glands immediately associated with the sting, there are two masses of glandular cells lying against the inner surfaces of the supper parts of the quadrate plates. Each cell opens by an individual duct into a pocket of the membrane between the quadrate plate and the overlapping edge of the spiracle plate above it, so that the secretion would appear to be discharged on the outer surface of the quadrate plate. These glands have been termed "lubricating glands," but it is not clear what parts of the sting mechanism might be lubricated by their secretion.

The Mechanism of the Sting— When the sting goes into action, the basal apparatus swings downward and posteriorly on the articulations of the quadrate plates with the spiracle plates until it takes an almost vertical position, with the sheath lobes of the oblong plates directed upward (fig. 62 A, B). At the same time the shaft is depressed and the posterior swing of the supporting basal apparatus protracts it from the tip of the abdomen. The lancets now begin rapid, alternating back-and-forth movements on the stylet. The complete action of the sting thus involves three movements: first the downward and posterior swing of the basal structure, second the simultaneous depression of the shaft, and third the movement of the lancets. On retraction of the sting the basal sclerites return to the position of repose (A) and the shaft is replaced between the sheath lobes.

The posterior swing of the sting base on the spiracle plates of segment VIII was formerly attributed by the writer (1935) to pressure supposed to be engendered within the abdomen by contraction of the abdominal segments, but it has been shown by Rietschel

160

(1937) to be caused by an upward movement of the anterior part of the sternum of segment VII. Manipulation of a dead specimen will demonstrate that when the seventh sternum is depressed posteriorly, its broad anterior part, turning upward and backward, is pressed deeply into the abdomen against the base of the sting and forces the latter to swing posteriorly. It may be questioned, however, whether this movement of the sternum accounts entirely for the displacement of the sting; the spiracle plates have an elaborate musculature (fig. 55 B), and there is no evident reason for individual movements of the spiracle plates themselves. Two small muscles (*187, 188*) go from the upper angle of each spiracle plate to the corresponding quadrate plate, and a long muscle (*192*) extends from the posterior end of the spiracle plate to the triangular plate. These muscles might have some action in restoring the sting to the retracted position. The posterior movement of the basal apparatus does not protract the shaft of the sting.

The deflection of the sting shaft that accompanies the posterior swing of the basal apparatus is evidently effected by a pair of large, flat muscles arising posteriorly on the inner faces of the oblong plates (figs. 61 C, 62 C, *197*) that extend anteriorly and converge over the base of the bulb to be attached on the furcula (*Frc*). The tension of these muscles on the furcula pulls on the base of the bulb (fig. 62 D), with the result that the entire shaft is turned downward (E) on the flexible connections (*h*) of the base of the bulb with the second rami (*2r*). In the retracted position of the sting (fig. 62 A) the shaft is turned upward and ensheathed between the oblong plates and their distal appendages (*Sh*); it is held in this position by a pair of muscles stretched between the second rami and the base of the bulb (figs. 61 D, 62 C, *196*). The insertions of these muscles are sufficiently beyond the hinge points of the bulb on the rami (*h*) to give them a short leverage on the base of the shaft.

The movements of the lancets on the stylet accomplish the penetration of the tip of the sting into the victim and the injection of the poison into the wound. They are produced by two pairs of large antagonistic muscles in the basal apparatus that activate primarily the quadrate plates and secondarily the triangular plates and the attached lancets. The muscles of one pair consist of two large bundles of fibers on each side (figs. 61 C, 62 C, *198*) that arise

161

posteriorly, one laterally, the other mesally, on the apodeme of the quadrate plate and are inserted on the anterior end of the oblong plate. The smaller muscles of the second pair (*199*) arise by broad

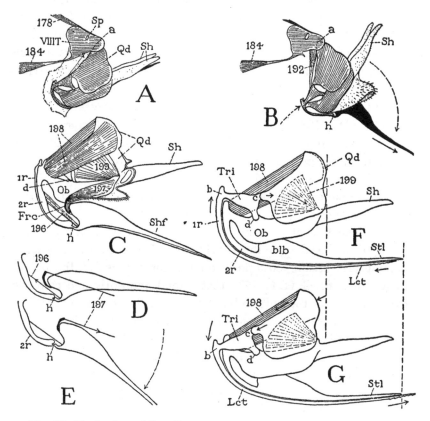

Fig. 62. Mechanism of the sting.

A, diagram of sting in retracted position, suspended from spiracle plates of segment VIII. B, the sting protracted. C, shaft of sting (*Shf*) and right half of basal apparatus with muscles, mesal. D, shaft of sting held in retracted position by muscle *196*. E, shaft protracted by muscle *197*. F, diagram showing muscles of oblong plate, lancet retracted by muscle *199*. G, same, lancet protracted by muscle *198*.

Alphabetical lettering and muscles are explained in text. For explanation of abbreviations see pages 166–167.

bases on the inner surfaces of the quadrate plates and are inserted posteriorly on the posterior ends of the oblong plates. The alternate contraction of these two opposing muscles on each side pulls the corresponding quadrate plate first forward and then backward relative to the oblong plate, and the movements of the quadrate plate are transmitted to the triangular plate, which rocks on its ventral articulation (fig. 62 F, G, *d*) with the oblong plate. The movements of the triangular plate in turn are communicated to the corresponding ramus of the lancet (F, *lr*), which finally gives the lancet itself a back-and-forth movement on the underside of the stylet.

The mechanism of the lancet movements is shown diagrammatically at F and G of figure 62. At F the quadrate plate is pulled back by muscle *199*, the triangular plate is rotated upward on the oblong plate, and the lancet (*Lct*) is thereby retracted. At G the quadrate plate is pulled forward by muscle *198*, the triangular plate is depressed, and the lancet is protracted. The muscles on the two sides of the sting work alternately, so that the lancets move simultaneously in opposite directions. In a live bee, as may be seen in a freshly extracted sting, the movements of the quadrate plates appear as rapid vibrations. Although the quadrate plates are parts of the tergum of segment IX, they are able to vibrate in this manner because of the membranization of the median part of the dorsum of their segment.

When a bee stings, the initial insertion of the tip of the outthrust sting into the skin of the victim is probably effected by a quick movement of the deflected end of the abdomen, but the subsequent deeper penetration is the result of the successive alternating movements of the lancets. Each lancet after each thrust holds the sting in place by means of the recurved barbs on its distal end, while the other lancet overreaches the first and holds the sting in a deeper position. The tension of the retractor muscles, now, instead of extracting the sting, probably depresses the anterior ends of the oblong plates and thus restores the triangular plates to a relative position in which the protractors again become effective. By the same action the stylet is enabled to follow the lancets into the wound. The point of the sting thus automatically goes deeper and deeper by the action of the movable lancets and will continue to do so even when the sting is detached from the body of the bee. At the same time the venom in the poison sac is poured into the bulb of the lancet and is driven through the poison canal by the action of the valvular lobes on the bases of the lancets. The poison escapes through a ventral cleft between the lancets near the tip of the sting.

The basal apparatus of the sting has only a delicate membranous

163

connection with the walls of the sting chamber. A very gentle pull on the tip of the sting, therefore, suffices to remove the sting from the chamber, and when extracted it brings with it the quadrate plates of the ninth segment, the poison glands, the proctiger, and the terminal parts of the alimentary canal. In the same way these parts may tear out from the living bee as the latter hurriedly leaves her victim after stinging. It is noted by Rietschel (1937) that the very thin integument that connects the sting with the spiracle plates constitutes a preformed breaking place, and Rietschel, therefore, regards the loss of the sting as a case of true autotomy. The advantage gained, of course, lies in the fact that the extracted stinging apparatus left in the skin of the recipient keeps on automatically pumping in poison, and thus gives a more effective dosage than the bee could otherwise deliver before being brushed off. Presumably this advantage compensates for the resulting disability and eventual death of the stinger. The self-sacrifice on the part of the bee, however, occurs only when the victim is a human being or other thick-skinned animal.

The sting of the queen (fig. 63 C) differs from that of the worker in several respects. The plates of the basal apparatus are not of the same shape or proportional sizes as in the worker and are more firmly attached within the sting chamber. The queen uses her sting only on rival queens, and obviously could not suffer the consequences of losing it. The shaft of her sting is decurved beyond the base, the lancets have fewer and smaller barbs than those of the worker's sting, but the poison glands are well developed (fig. 104), and the poison sac is very large.

A study of the bee's sting thus shows that the organ is a highly intricate mechanism perfectly adapted in every detail to the work it has to do; probably no human genius could devise an improvement. Perhaps an entirely different kind of instrument could be just as effective, but the sting of the bee has been evolved from an ovipositor, and hence its fundamental structure must be that of an egg-laying apparatus. That the sting is an elaborated ovipositor is shown in the identity of the origin and development of the two organs.

The Development of the Sting— The rudiments of the bee's sting are first to be seen in an early stage of the pupa, still within the larval cuticle (fig. 63 A), as three pairs of small, budlike outgrowths,

164

known as *gonapophyses*, one pair (*1G*) on the venter of the eighth segment just behind the genital opening (*Gpr*), the other two pairs (*2G, 3G*) on the ninth segment. At a later stage (B) the rudiments have elongated into slender fingerlike processes. Those from segment VIII extend beneath the others and will become the lancets

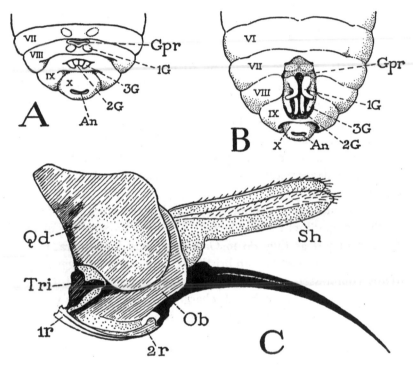

Fig. 63. Early developmental stages of the sting of the worker, and the sting of the queen.

A, early stage of worker pupa with rudiments of first gonapophyses on segment VIII of abdomen and rudiments of two pairs on segment IX. B, later stage of same, gonapophyses lengthened and assembled in position of parts of sting. C, sting of adult queen.

For explanation of abbreviations see pages 166–167.

of the mature sting. The median pair on segment IX will unite to form the stylet, and those of the outer pair will form the oblong plates and the sheath lobes. The triangular plates of the sting are probably differentiated from the bases of the first gonapophyses on segment VIII.

An ovipositor develops in the same way as does the sting, and works in much the same manner, but the mechanism necessary for discharging eggs is in general simpler than that for injecting poison. In the more general terminology of entomology for the parts of an ovipositor, the six elements derived from the primary rudiments are called *valvulae*. The lancets of the bee's sting are the *first valvulae*, the stylet is the united *second*, or *median*, *valvulae* of the ninth segment, and the sheath lobes of the sting are the *third valvulae*. The triangular plate of the sting is the *first valvifer*, the oblong plate the *second valvifer*. The triangular plate belongs to segment VIII, its single muscle (fig. 61 E, *192*) arises on the spiracle plate of the eighth tergum, but for mechanical reasons the triangular plate has acquired a secondary articulation on the quadrate tergal plate of segment IX. The oblong plate clearly belongs to segment IX because its muscles arise on the quadrate plate. Theoretically, the triangular plates and the quadrate plates represent the bases of paired appendages pertaining to the eighth and the ninth abdominal segments.

In the nonstinging members of the Hymenoptera the females use their ovipositors for egg laying. The eggs traverse the channel of the shaft just as they do in Orthoptera and Hemiptera, though in some parasitic species with bristlelike ovipositors the eggs may be greatly compressed and stretched in their transit through the narrow channel. The queen bee discharges her eggs directly from the genital opening before the base of the sting.

Explanation of Abbreviations on Figures 51–63

Ac, antecosta.
AGld, poison gland of sting.
Amp, ampulla.
An, anus.
ap, apodeme of tergum IX of drone.
Ap, apodeme of quadrate plate of sting.

Bgld, accessory gland of sting.
blb, bulb of stylet of sting.

Can, canal on tergum VII of worker.
cmcl, compressor muscle.
CxP, pleural coxal process.

Endst$_{2+3}$, composite endosternum of mesothorax and metathorax.
Epd, epidermis.

Frc, furcula.
FtCls, fat cells.

G, gonapophysis; *1G*, first gonapophysis (lancet), *2G*, second gonapophysis (stylet), *3G*, third gonapophysis (sheath lobe of sting).
gd, genital disc of larva.
Gpr, gonopore.

Lct, lancet of sting.
lp, lamina parameralis.
Lsp, lamina spiracularis of segment VIII.

Mb, intersegmental membrane.
Mir, "mirror" under wax gland.
Mmr, mesomere.

Ob, oblong plate of sting.
Oen, oenocytes.

pc, poison canal of sting.
2Ph, postphragma of mesothorax.
PhB, phallobase.
phL, primary phallic lobe.
Phtr, phallotreme.
pmcl, protractor muscle.
Pmr, paramere.
PsnSc, poison sac of sting.
Ptgr, proctiger.
pv, penis valve (mesomere).

Qd, quadrate plate of sting.

1r, 2r, first and second basal rami of sting.
Rect, rectum.
rmcl, retractor muscle.

S, sternum.
IS, sternum of propodeum.
SA, arm of endosternum.
Sh, sheath lobe of sting.
SntGld, scent gland.
Sp, spiracle.
Stl, stylet of sting.
Stng, shaft of sting.

T, tergum.
IT, tergum of propodeum.
Tri, triangular plate of sting.

IX.V, membranous venter of segment IX.
vDph, ventral diaphragm.
Vlv, valve on lancet of sting.

WxGld, wax gland.

IX

THE ALIMENTARY CANAL

THE alimentary canal of an insect consists of three parts, which are separate in their embryonic origin (fig. 64 A) but become united to form in the adult (B) a continuous food tract extending through the body. The primary middle section, known as the *mesenteron* (A, *Ment*), is of endodermal origin and will be the functional stomach, or *ventriculus* (B, *Vent*), of the adult. The anterior section, or *stomodaeum* (A, *Stom*), opening from the mouth, and the posterior section, or *proctodaeum* (*Proc*), opening at the anus, are secondary ingrowths of the ectoderm, and in the embryo they are closed at their inner ends. In the adult (B) the stomodaeum and the proctodaeum are not only open into the ventriculus, but each is commonly differentiated into several parts. The stomodaeum includes a *pharynx* (*Phy*), a narrow *oesophagus* (*Oe*), a baglike *crop* (*Cr*), and a short *proventriculus* (*Prvent*). The true mouth of the insect (*Mth'*), as already explained, is the opening from the preoral cibarium (*Cb*) into the stomodaeal pharynx (*Phy*). The proctodaeum consists of a usually slender *anterior intestine* (*AInt*) and of a posterior saclike *rectum* (*Rect*); its anterior end may be expanded into a small *pylorus* (*Pyl*) against the end of the ventriculus. Opening into the pylorus are a number of excretory *Malpighian tubules* (*Mal*), from two to a hundred or more in different insects.

The names given above to the parts of the insect alimentary canal are those that will be used here in describing the food tract of the bee. For the benefit of the student the following synonyms may be listed: STOMODAEUM, fore-gut, *Vorderdarm;* CROP, ingluvies, *jabot,* honey stomach, honey sac; PROVENTRICULUS, *Zwischendarm, Ven-*

168

tiltrichter; MESENTERON, ventriculus, stomach, mid-gut, *Mittledarm;* PROCTODAEUM, hind-gut, *Hinterdarm, Enddarm;* ANTERIOR INTESTINE, small intestine, *Dünndarm;* RECTUM, large intestine, *Kotblase.* The anterior intestine is not always "small" or "slender."

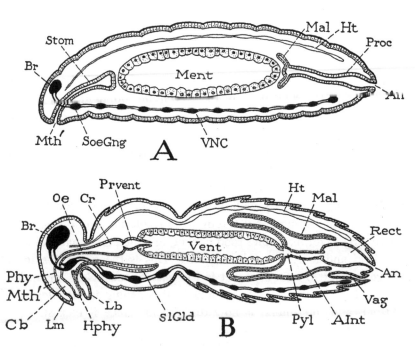

Fig. 64. Composition of the insect alimentary canal and differentiation of its parts.

A, diagrammatic section of an embryo showing the three primary component parts of the alimentary canal; the endodermal mesenteron and extodermal stomodaeum and proctodaeum. B, the adult insect and fully developed alimentary canal.

For explanation of abbreviations see pages 199–200.

EVOLUTION AND EMBRYOGENY OF THE FOOD TRACT

The alimentary canal of an insect such as the bee cannot be described as a single structure, because during the life of each individual it goes through a succession of changes from its inception in the embryo to its final form in the adult. First it must be adapted to serve the embryo shut up in an eggshell and supplied with food in the form of yolk stored in the egg; then it is recon-

169

structed in adaptation to the needs of the larva confined in a cell of the honeycomb and fed by adult attendants; finally, in the pupa the whole food tract undergoes a radical transformation in anticipation of the very different functions it will have to serve in the free-living adult, and at the same time it makes certain temporary adjustments for the pupa itself. Necessarily, therefore, the insect in its ontogenetic development is forced to depart from the phylogenetic method of forming an alimentary canal; and to understand the embryonic processes we must attempt to understand the probable evolution of the food tract in the free-living progenitors of the insects, which took their food from the outside by way of an open mouth.

From the development of some of the simpler invertebrates we can infer that at a very primitive stage of its evolution the animal had only a surface layer of cells (fig. 65 A) and absorbed its nourishment through the body wall. Later, however, a more efficient manner of feeding came about through the infolding of one side of the body (B) to form an inner food pocket, in which food material could be held and more leisurely digested. This is the *gastrula* stage of development; the food pocket is the *gastrocoele* (*Gcl*), or *archenteron*, i.e., the most ancient stomach, its opening is the *blastopore* (*Bpr*), and the animal is now diploblastic since it consists of an outer *ectoderm* (*Ecd*) and an inner *endoderm* (*End*).

In the gastrula stage the blastopore is commonly a hole at the posterior end of the embryo, but the development of the arthropods suggests that the blastopore in the progenitors of these animals took the form of a long slit on the undersurface of the body (fig. 65 D, *Bpr*), which became closed along the middle, leaving an opening at the anterior end as the mouth (*Mth*) and one at the posterior end as the anus (*An*). An embryonic blastopore of this kind, in fact, is typically formed in some of the Onychophora, which animals are certainly related to the primitive arthropods. The animal thus acquires a stomach extending through the length of the body (F) with an anterior mouth and a posterior anus opening directly into it. Subsequently both the mouth and the anus are carried inward by tubular ingrowths of the ectoderm (G) that become, respectively, the stomodaeum (*Stom*) and the proctodaeum (*Proc*). The alimentary canal now has the typical three-part structure. The original archenteron has become the mesenteron (*Ment*), and the

170

definitive mouth (Mth') and the definitive anus (An') are the external openings of the stomodaeum and the proctodaeum.

Finally, we must note that in a triploblastic animal the mesoderm is usually associated in its origin with the endoderm, being formed

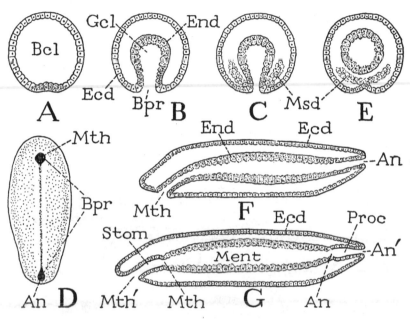

Fig. 65. Probable evolution of the alimentary canal in the free-living progenitors of the arthropods.

A, section of an animal in the blastula stage. B, same in gastrula stage, one side turned in to form a food pocket, the archenteron, or gastrocoele (Gcl), the wall of which is the endoderm (End), the opening the blastopore (Bpr); outer wall of gastrula is now the ectoderm (Ecd). C, same, mesoderm bands (Msd) proliferated from lips of blastopore. D, undersurface of an elongate animal with blastopore closed except for an anterior mouth (Mth) and a posterior anus (An). E, cross section of same, gastrocoele now tubular. F, same as D in longitudinal section. G, later stage of same, primary mouth (Mth) and anus (An) carried inward with formation of ectodermal stomodaeum ($Stom$) and proctodaeum ($Proc$), establishing a permanent secondary mouth (Mth') and secondary anus (An'); the gastrocoele now a mesenteron ($Ment$).

either from the lips of the blastopore (fig. 65 C, Msd) or as part of a common endoderm-mesoderm rudiment. When the blastopore closes, therefore, the mesoderm (E, Msd) appears as a sheet of tissue below the alimentary canal.

Now let us see how the embryo of the honey bee modifies its

171

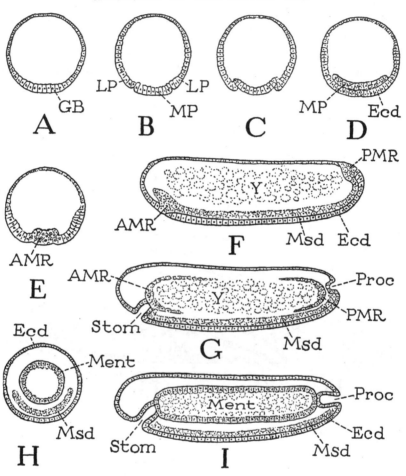

Fig. 66. The embryonic method of forming the alimentary canal in the honey bee (diagrams based on Nelson, 1915).

A, cross section of egg in blastula stage, the germ band (*GB*) a thickening of lower side of blastoderm. B, germ band differentiated into lateral plates (*LP*) and a ventral median plate (*MP*). C, lateral plates converging beneath median plate. D, median plate entirely enclosed, becomes mesoderm and endoderm. E, cross section of anterior part of egg, showing anterior mesenteron rudiment (*AMR*) still continuous with ectoderm. F, lengthwise section of egg, showing anterior mesenteron rudiment (*AMR*) and posterior mesenteron rudiment (*PMR*) at opposite ends of mesoderm. G, same, mesenteron rudiments enclosing the yolk (*Y*) to form the mesenteron; stomodaeum (*Stom*) and proctodaeum (*Proc*) growing inward from ectoderm. H, cross section of embryo with completed mesenteron. I, lengthwise section of same stage as H.

phylogenetic inheritance in the formation of its alimentary canal, and for this purpose we may follow the account by Nelson (1915) of the embryogeny of the bee. At an early stage of development (fig. 66 A) the embryo is represented by a ventral thickening of the blastoderm known as the *germ band* (*GB*). The germ band spreads dorsally on each side of the egg (B) and becomes differentiated into two *lateral plates* (*LP*) and a ventral *middle plate* (*MP*). Next the middle plate sinks into the yolk (C), and its margins are under-lapped by the edges of the lateral plates, which finally come to-gether (D) to complete the ventral ectoderm (*Ecd*); the middle plate thus becomes an internal layer. Most of the middle plate is mesodermal (F, *Msd*), but masses of cells at its opposite ends (*AMR, PMR*) are endodermal since they will form the mesenteron of the future insect, for which reason they are termed the *anterior* and the *posterior mesenteron rudiments*. The mesenteron rudiments at this stage are not entirely covered by the ectoderm as is the mesoderm (E, *AMR*). The middle plate of the embryo thus con-sists of prospective mesoderm and prospective endoderm, and it is evident that its sinking into the body represents an infolding along the line of a primitive median blastopore, which is closed over by the ectoderm except at the two ends, and it is here that the stomo-daeum and proctodaeum will later be formed (G). The mesenteron rudiments, therefore, represent the two ends of the blastopore, and it is to be presumed that originally the endoderm extended along the whole length of the blastopore; in some insects a few endoderm cells are distributed over the mesoderm between the two end masses.

The mesenteron rudiments of the bee, then, are clearly mere anterior and posterior remnants of a former endoderm invaginated along the line of a median ventral blastopore. It is the business of the embryonic endoderm to enclose the yolk in a stomach wall. This could not be done by a stomach formed intact in the ancestral man-ner by invagination, but it can readily be accomplished by a dis-rupted endoderm. The mesenteron rudiments in their growth first form cuplike enclosures of the two ends of the yolk (fig. 66 G) and then send out arms toward each other over the yolk, from which sheets of cells extend downward on the sides until finally the yolk is enveloped in a closed bag, which is the future mesenteron (H, I, *Ment*).

During the formation of the mesenteron there appear at the ends

173

of the embryo, where the endoderm was not covered by the ectoderm, shallow ingrowths of the ectoderm that will become the
stomodaeum and the proctodaeum of the larva (fig. 66 G, *Stom*,
Proc). During the life of the embryo, however, these ingrowths do
not open into the mesenteron (fig. 67 A, *Stom*), since the embryo

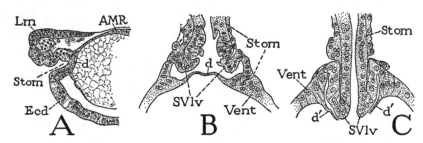

Fig. 67. Formation of the stomodaeum of the honey bee and its later opening
into the ventriculus (from Nelson, 1915).

A, stomodaeal ingrowth of embryo 48–50 hours old. B, junction of stomodaeum with ventriculus at end of embryonic life and formation of stomodaeal
valve. C, later stage of same, stomodaeal valve opened through the ventricular
partition.

d, partition between stomodaeum and ventriculus; d', remnant of broken partition. For explanation of abbreviations see pages 199–200.

could not use a mouth if it had one and an anal opening from the
stomach would be an extreme inconvenience.

The embryonic development of the honey bee in the egg occupies
three days, during which time the embryo subsists on the original
supply of yolk. At the end of this period the insect must be equipped
for its coming life as a larva, when it must ingest the food placed in
its cell by the nurse bees. Now, therefore, for the first time, the
stomodaeum opens into the mesenteron. As described by Nelson
(1915), the partition that heretofore has closed the anterior end of
the mesenteron, or ventriculus, is reduced to a thin membrane
(fig. 67 B, d). The epithelium of the inner end of the stomodaeum
forms a circular fold (*SVlv*), which enlarges posteriorly against the
partition, breaks through the latter, and extends into the lumen of
the ventriculus as a double-walled tube known as the *stomodaeal
valve* (C, *SVlv*), which now gives an open passageway from the
stomodaeum into the stomach. The proctodaeum, on the other hand,
remains closed as in the embryo until the end of the larval period
to prevent the larva from contaminating the food in its cell. Larvae

174

of insects that live in the open have a complete alimentary canal from the time of hatching. Hence we see again how the course of development can be modified to suit each particular kind of insect according to the way it lives. In short, ontogenetic development is an adaptation of old ways to new conditions and varies as conditions vary.

THE LARVAL ALIMENTARY CANAL

The alimentary canal of the honey-bee larva is a relatively simple structure (fig. 68 A), consisting of a short, slender stomodaeum (*Stom*), a long, cylindrical mesenteron, or ventriculus (*Vent*), ex-

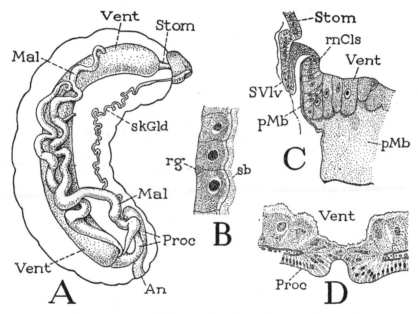

Fig. 68. Alimentary canal of honey-bee larva. (A, B, C from Nelson, 1924; D from Rengel, 1903).

A, entire alimentary canal of mature larva. B, part of ventricular epithelium of old larva. C, junction of stomodaeum and ventriculus, showing half of stomodaeal valve, and peritrophic membrane. D, union of ventriculus and proctodaeum in a feeding larva.

For explanation of abbreviations see pages 199–200.

tending through most of the length of the body, and a narrow proctodaeum (*Proc*), which makes a forward turn ventrally before going back to the anus (*An*). During the life of the larva the stomodaeal valve (fig. 69 A, *SVlv*) remains open into the ventriculus; but posteriorly the ventriculus is shut off from the proctodaeum by a thick partition (fig. 68 D) in which are united the opposing ventricular and proctodaeal walls. Four large Malpighian tubules (A, *Mal*) arise between the ventriculus and the proctodaeum and go forward, two on each side, along the ventriculus as far as the thorax. Their tapering central ends are closed and are inserted between the two layers of the partition between the ventriculus and the proctodaeum (fig. 76 B, *Mal*).

The walls of the larval ventriculus are shown by Nelson (1924) to be a simple epithelium of cubical cells (fig. 68 B), on the inner surface of which is a distinct striated border (*sb*). Here and there, wedged between the bases of the epithelial cells next the basement membrane, are triangular groups of minute cells (*rg*), which are the elements that will regenerate a new epithelium in the pupa. Externally the ventriculus is covered by a delicate meshwork of muscle fibers, which, according to Dobrovsky (1951), consists of the usual outer longitudinal and inner circular fibers of the ventriculus but includes in addition a third innermost layer of small longitudinal fibers, which probably are branches of the circular fibers.

A peritrophic membrane lines the epithelium of the larval ventriculus throughout its length; it is described by Nelson (1924) as a homogeneous layer of apparently gelatinous consistency two or three times as thick as the epithelium itself (fig. 68 C, *pMb*). Though this layer is closely applied to the epithelium, Nelson believed it to be secreted by a ring of special cells (*rnCls*) in the anterior end of the ventriculus around the base of the stomodaeal valve (*SVlv*). From these cells, he says, "thin streams of secretion can plainly be seen running caudad and joining with the principal mass of the peritrophic membrane in the midintestine." According to Kusmenko (1940), on the other hand, there are in the honey-bee larva two distinct peritrophic membranes of different origin. One is the thick gelatinous membrane described by Nelson, which Kusmenko says is formed from the general surface of the ventricular epithelium; the other is produced by the ring of special cells in the anterior end of the ventriculus, the secretion of which is granular and flows backward as an inner membrane to the end of the ventriculus. The two membranes, however, may unite and thus give the appearance of a single thick membrane such as that described by Nelson.

After about the fifth day of its life the worker larva, now in its fifth instar, is sealed within its cell. Then, according to Bertholf (1925), it gorges itself on the remaining food and attains its maturity at the end of the sixth day. Its feeding stage is now over; the ventriculus and the Malpighian tubules are distended with accumulated waste matter. Soon, however, the septum between the ven-

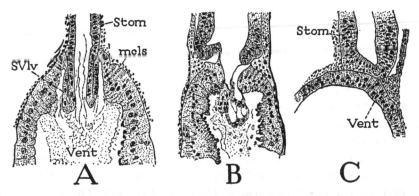

Fig. 69. Stages in the closure of the larval stomodaeum into the ventriculus after the discharge of the larval feces, about 8 days after hatching (from Christa Evenius, 1926).

A, stomodaeal valve still open. B, valve retracted, walls of ventriculus coming together. C, closure fully established.

For explanation of abbreviations see pages 199–200.

triculus and the proctodaeum (fig. 76 B) is disrupted and the end of the ventriculus extends into the proctodaeum as an open tubular conduit, or ventricular valve (C, *VVlv*), through which the feces are discharged. At the same time the hitherto closed inner ends of the Malpighian tubules (B, *Mal*) open into the proctodaeum, and their contents are voided with the feces.

METAMORPHOSIS OF THE ALIMENTARY CANAL

With the close of the larval period, the processes of metamorphosis begin at once, but the last larval cuticle is not shed until the eighth day, so that for two days the propupa is still enclosed in the larval skin. During the early part of the propupal period the stomodaeum, which during the larval life has been freely open into the ventriculus, is shut off from the latter by a newly formed partition. Likewise, at the end of this period the opening of the ventriculus into the proctodaeum is again closed. The closure of the stomodaeum

177

is described in detail by C. Evenius (1926) and more briefly by Dobrovsky (1951). First the larval stomodaeal valve (fig. 69 A, SVlv) becomes shortened and retracted into the stomodaeum (B). Then again it lengthens, but now its margins come together and unite, forming a wall across the end of the stomodaeal lumen (C). At the same time a thick epithelial fold forms in the anterior end of the ventriculus, the edges of which unite in a central plug (B). The cells of the plug, however, become vacuolated and go into histolysis, while the lateral cells of the fold come together forming a second wall of ventricular cells, which is closely applied to the stomodaeal wall. The stomodaeum and the ventriculus of the pupa are thus separated by a thick, double-walled partition (C). The closure of the ventriculus into the proctodaeum is said by Oertel (1930) to be effected by a plug of tissue derived from the adjacent epithelial cells of the ventriculus.

The metamorphic processes that take place in the pupa rapidly transform the simple food tract of the larva (fig. 68 A) into the diversified alimentary canal of the adult (fig. 70). The successive stages of the process are graphically illustrated by Dobrovsky (1951) in a series of 12 drawings showing the changes in the form of the canal and in the position of its parts in the body of the bee. The stomodaeum greatly elongates and becomes differentiated into an oesophagus, a crop ("honey stomach" of the bee), and a proventriculus; the ventriculus takes the form of a long, cylindrical sac looped upon itself; the proctodaeum becomes differentiated into a slender anterior intestine and a large saclike posterior intestine, or rectum. During these changes all parts of the alimentary canal except the oesophagus are crowded back into the abdomen, thus giving over the thorax to the developing leg and wing muscles.

Early in the transformation period the Malpighian tubules of the larva disintegrate and go into complete dissolution, but before they disappear the tubules of the adult begin to form as a ring of small outgrowths around the end of the pyloric region of the proctodaeum (fig. 76 D, iMal). In the bee, therefore, the numerous slender tubules of the imago are quite distinct in their origin from the larval tubules.

The principal sources of our information on the pupal metamorphosis of the alimentary canal in the honey bee are the papers by Anglas (1901), Oertel (1930), Lotmar (1945), and Dobrovsky (1951). In the reconstruction of the alimentary canal the gross external changes are accompanied by important activities of the epithelial cells, particularly in the ventriculus. The ventriculus of the larva, as Lotmar points out, is functionally a larval organ, and hence its epithelium must be replaced with a new epithelium suitable to the pupa and the adult. The stomodaeum and the proctodaeum, on the other hand, are of relatively little importance in the larva, and they are merely enlarged and elaborated in structure to serve the adult.

Changes begin in the ventriculus in the propupal stage. The larval epithelial cells degenerate and are thrown out into the ventricular lumen; only the basement membrane remains intact. At the same time the cells of the regenerative centers actively multiply and spread out over the basement membrane to form a new pupal epithelium replacing that of the larva. The process of replacement, Dobrovsky says, begins anteriorly and proceeds posteriorly, but since the basement membrane remains intact, the insect is at no time actually without a ventriculus. The pupal epithelium, however, is itself temporary. At about the fifth day of the pupal period, according to Lotmar, a new activity begins in the regenerative centers, the cells of which again multiply and spread out, this time to form a definitive imaginal epithelium that now replaces the pupal epithelium. The imaginal epithelium, Lotmar says, differs in appearance and in staining properties from the pupal epithelium and is completed at the time of the last moult, about a day and a half before the emergence of the young bee. A reason for the formation of a temporary pupal epithelium is not evident, since the observations of Lotmar do not support the idea that the discharged larval cells are digested and absorbed by the pupal cells to be used over again as food for the growing tissues. At the time the larval cells are discharged, she says, the pupal epithelium is not yet fully formed and is not in a functional condition.

The thick, gelatinous peritrophic membrane of the larval ventriculus, according to Dobrovsky, disappears with the dissolution of the larval cells, and a new, thinner membrane is formed shortly after the larva ceases to feed. This membrane encloses the entire undigested material that has accumulated in the larval ventriculus and is carried with the food refuse into the proctodaeum when the latter is opened. No other peritrophic membrane, Dobrovsky says, is formed until the end of the pupal period, when the definitive imaginal epithelium of the ventriculus is nearing completion. The discharged larval epithelial cells, therefore, fall into the lumen of the ventriculus while no peritrophic membrane is present.

The strong muscular coat of the larval ventriculus is said by Dobrovsky to be carried over into the propupal stage, but it disappears in the young pupa, and in its place are found only small myoblasts and a few slender fibers. Not until the fourth day of the pupa is the ventriculus again provided with a muscular sheath of longitudinal and circular muscles.

The metamorphic changes that take place in the ectodermal parts of the alimentary canal are gradual and merely reconstructive. In both the stomodaeum and the proctodaeum active division takes place in the cells of the epithelium at the inner ends of the tubes, but in the bee regeneration does not proceed from definite "imaginal rings" as in some other insects. The zone of proliferation extends forward in the stomodaeum, backward in the proctodaeum, and replaces the larval epithelia. According to Anglas, the new cells absorb the old cells as they advance and take their places. Reconstruction in these two ectodermal parts of the alimentary canal, therefore, takes place in the same manner as in the ectodermal body wall, but it involves great changes of form and structure in both the stomodaeum and the proctodaeum.

By the end of the pupal period the alimentary canal is fully reconstructed in the adult form, but before it can become a functional organ the partitions that have closed the two ends of the ventriculus during the pupal stage must be removed. The opening of the stomodaeum into the ventriculus takes place about two days before the pupal moult. The process, which includes the formation of the definitive stomodeal valve, has been described by Metzer (1910), J. Evenius (1925), and Dobrovsky (1951), and is here illustrated with drawings copied from Evenius (fig. 72 A, B). While the ventriculus is still shut off from the stomodaeum by a double-walled partition (A), there is formed in the posterior part of the proventriculus a circular fold of the epithelium extending forward in the proventricular lumen. This fold (SVlv) is the beginning of the stomodaeal valve; it everts toward the ventriculus as the anterior wall of the partition becomes perforated by an opening. Then the central cells of the posterior wall become vacuolated and go into a state of dissolution, so that an opening is thus formed from the

180

stomodaeum into the ventriculus. The valvular fold of the stomodaeum now everts into the ventriculus (B) and here forms the
long, double-walled tubular extension of the stomodaeal epithelium
known as the stomodaeal valve (C, SVlv). Finally, at the end of
the pupal period, the ventriculus opens into the proctodaeum. The
"tissue plug" that has closed the ventriculus during the entire pupal
stage is said by Oertel (1930) to break loose from the ventriculus
and to move into the intestine. The alimentary canal is thus at last
ready for active service in the adult bee, which now emerges from
its cell.

THE ALIMENTARY CANAL OF THE ADULT BEE

During the pupal transformation the alimentary canal has been
so greatly enlarged and elaborated in its structure that in its final
form in the adult bee (fig. 70) it has little resemblance to the simple
food tract of the larva (fig. 68 A). It now serves not only for the
intake and digestion of food and the discharge of waste products,
but also as a carrier of nectar and honey. The parts of the mature
alimentary canal of the bee conform entirely with those shown in the
diagram at B of figure 64; so we may proceed at once with a descrip-
tion of them.

The Pharynx— The pharynx is the first part of the stomodaeum.
In the bee, as already explained, its walls and its lumen are continuous with those of the preoral cibarium, so that the two parts
appear as a single sac in the head (fig. 17 B, C). Most students of
bee anatomy, including the writer (1925), therefore, have errone-
ously described this entire sac as the "pharynx," but this error need
no longer be perpetuated.

The Oesophagus— The oesophagus is a simple, slender tube (fig.
70, Oe) continued from the narrowed upper end of the pharynx
through the thorax into the abdomen, where it expands into the
large saclike crop (Cr). Its walls, which are lined with a thick
cuticular intima, are thrown into numerous circular folds allowing
of expansion (fig. 72 C, Oe). On the outside is a strong muscular
layer of external circular fibers and inner longitudinal fibers.

The Crop— The crop, or ingluvies, of the bee's alimentary canal
is commonly known in bee literature as the "honey stomach," but
it is not a stomach in any physiological sense. Its principal function
is that of a carrier of nectar being transported to the hive for conver-

181

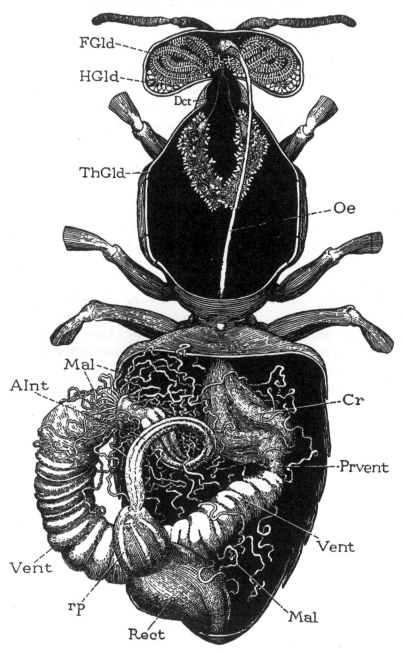

Fig. 70. Alimentary canal and glands of the head and thorax of a worker bee, dorsal.

For explanation of abbreviations see pages 199–200.

sion into honey. Anatomically the crop is merely an enlargement of the oesophagus (fig. 71 A), and its walls have the same structure as those of the oesophagus (fig. 72 C, *Cr*). When the crop of the worker is filled with nectar, it becomes a great balloon-shaped bag with thin, tense walls, but when empty it collapses to a small flabby

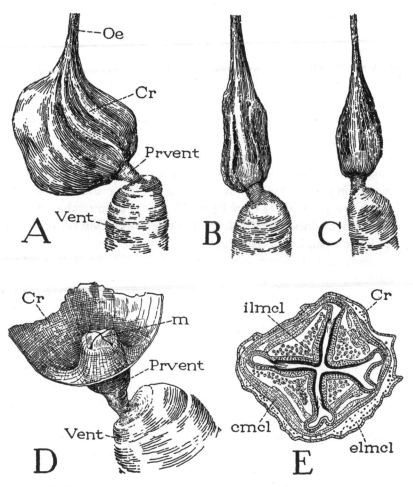

Fig. 71. The crop (honey stomach) and proventriculus.

A, crop, proventriculus, and upper end of ventriculus of a worker. B, same of a queen. C, same of a drone. D, wall of crop mostly cut away, exposing mouth of proventriculus projecting into the crop. E, cross section of proventriculus (from Trappmann, 1923).

m, mouth of proventriculus. For explanation of abbreviations see pages 199–200.

pouch. The organ in the queen and the drone (fig. 71 B, C) is much slenderer than that in the worker (A).

The Proventriculus— The proventriculus (*gésier, Zwischendarm, Ventiltrichter*) is a short section of the stomodaeum intervening between the crop and the ventriculus, but, since its anterior end is invaginated into the crop (fig. 71 D), its exposed part appears as a neck supporting the crop on the ventriculus (A, *Prvent*). If the crop is cut open (D), there is seen a thick elevation of its posterior wall with four triangular lips closing an X-shaped opening on its truncate summit. This opening (*m*) is the mouth of the proventriculus, which is pushed into the rear wall of the crop. The lips are the ends of four thick triangular folds of the proventricular wall (E), which are lined with a dense cuticular intima. Each fold contains a large bundle of longitudinal muscle fibers (*ilmcl*), and the whole organ is surrounded by a thick sheath of circular fibers (*cmcl*). The lips are armed with groups of spines directed into the central lumen. A longitudinal section of the proventriculus of a queen bee is shown at C of figure 72. The proventriculus (*Prvent*) of the queen does not differ from that of a worker, but the crop (*Cr*) is slenderer and more symmetrical. The thick bundles of inner longitudinal muscle fibers (*lmcl*) are seen in two opposite lip folds, enclosed by the circular fibers (*cmcl*). Trappmann (1923) reports the presence of outer longitudinal muscle fibers (fig. 71 E, *elmcl*) not shown in figure 72 C.

The epithelium of the posterior end of the proventriculus is produced into the ventricular lumen as the long, double-walled tubular fold (fig. 72 C, *SVlv*) here termed the stomodeal valve, but known also as the oesophageal, proventricular, and cardiac valve and the *Ventilschlauch*. Of these names, "stomodaeal valve" is more generally applicable. A similar structure is present in nearly all insects and probably serves to prevent the ordinary movements of the ventriculus from forcing food back into the proventriculus, while at the same time it offers a free passage to food entering the stomach.

The function of the proventriculus in the worker honey bee is to regulate the entrance of food from the crop into the ventriculus and to retain in the crop the nectar to be taken to the hive. The opening into the proventriculus is the functional mouth of the stomach (*Magenmund*). If the crop in a freshly killed bee is cut open, the proventricular mouth may be seen still in action; the four lips, each armed with slender recurved spines, open wide with a quivering motion and then roll together tightly and sink into the mouth. Whitcomb and Wilson (1929) observed that when bees are fed pollen,

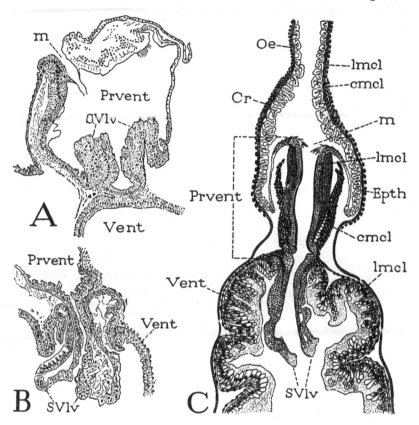

Fig. 72. The final opening of the stomodaeum into the ventriculus and the structure of the adult proventriculus and valve.

A, formation of stomodaeal valve in 15-day worker pupa (from J. Evenius, 1925). B, eversion of the valve into ventriculus in 18-day pupa (from J. Evenius, 1925). C, lengthwise section of crop, proventriculus, stomodaeal valve, and anterior end of ventriculus of a queen.

m, mouth of proventriculus. For explanation of abbreviations see pages 199–200.

the grains first collect at the mouth of the proventriculus. The whole organ now pushes into the crop with the mouth open, then the lips quickly close, draw backward, and bring with them pollen grains, which are passed on into the ventriculus. If the crop at the same time contains nectar or honey, very little is removed with the pollen.

Schreiner (1952) says that pollen is strained out from the liquid by the spines of the proventricular mouth and that the liquid is squeezed back into the crop. Bailey (1952) finds that a more normal action of the proventriculus is seen if it is observed through a "window" cut in the back of a bee. The lips of the proventriculus, he says, do not work together in the same rhythm; they snap open and close very rapidly but individually. The crop continually writhes and pulsates vigorously, keeping the contents well stirred and the pollen grains evenly distributed. The pollen is filtered off in compact masses, leaving the liquid behind in the crop. In the proventriculus the pollen grains are packed tightly together and the bolus is passed along into the ventriculus. In their passage through the ventriculus the pollen grains are not crushed or broken. The pollen mass, according to Bailey, reaches the posterior end of the ventriculus in from 5 to 20 minutes and remains here a varying length of time, 3 hours in a foraging bee and 12 hours or more in a brood-rearing bee, before it is passed on into the intestine.

The Ventriculus— The ventriculus is the functional stomach of the insect. In the honey bee it is a thick, cylindrical tube bent upon itself in a U-shaped loop (fig. 70, *Vent*) and is the largest part of the alimentary canal. Its surface is cut by numerous transverse constrictions that form deep internal folds. Examined under alcohol the ventriculus has an opaque white appearance, but as seen in a freshly killed bee it is of a brownish color with lighter rings corresponding with the constrictions. The color is due to the brown contents showing through the thinner walls between the folds.

The epithelial wall of the ventriculus begins at the base of the outer lamella of the stomodaeal valve (fig. 72 C, *SVlv*), which marks the end of the ectodermal part of the canal and the beginning of the endodermal stomach. The major transverse folds of the epithelium are formed by infoldings of the cell layer, the basement membrane, and the circular muscles, but between them are numerous minor folds and irregularities of the epithelium that do not involve the muscles. According to Trappmann (1923) and Weil (1935), the ventriculus has a connective tissue sheath outside the basement membrane. The surrounding muscles ordinarily observed are an external layer of longitudinal fibers (fig. 73 A, *lmcl*) and an inner layer of circular fibers (*cmcl*), but White (1918) described a third layer of inner longitudinal fibers (F, *ilmcl*) between the circular fibers and the basement membrane. Although Weil suggests that White mistook the connective tissue sheath for muscles, Morison (1928a) described and figured an inner layer of longitudinal muscles on the whole length of the ventriculus of the bee, the fibers of which branch and unite irregularly with one another, while some of them go to the muscles of the circular and outermost longitudinal layers.

A simple condition of the ventricular epithelium, which is probably a resting stage of the cells, is shown at B of figure 73. The epithelium is here thrown into minor folds around cuplike depressions, at the bottom of which are nests of regenerative cells (rg). Between the folds and covering their ends is a clear granular substance apparently of a gelatinous nature, which in some specimens forms a continuous layer over the entire epithelium (C). At the bottom of this layer is a very faint striated border of the cells (B, sb). Some writers, as Trappmann (1923), regard this whole covering layer as a product of the striated border, or rhabdorium. Weil (1935) says the rhabdorium is a plasmatic formation of the epithelial cells, which toward the lumen of the ventriculus may be thickened to a dense gelatinous sheath. From its intense blue coloring in Azan preparations, he concludes it is strongly impregnated with digestive secretion.

A more common condition of the ventricular epithelium is that seen at C of figure 73. Here the cells of the inner ends of the folds appear to be actively dividing and proliferating a large number of small nucleated cells into the covering mass. These liberated cells are most probably charged with digestive enzymes, since those in the containing matrix appear to be in all stages of disintegration. Again, at other places a different type of activity may be seen in the epithelial cells. The inner ends of some of the cells are extended in long necks containing clear spaces, evidently filled with liquid in life; in others the necks are constricted at their bases, while in still others the swollen ends have been cut off as free globules. This is the usual method of secretion discharge in the ventriculus of other insects, but both methods are described in the bee by Trappmann (1923), the second being well illustrated in figure 5 of his paper on the peritrophic membrane. The digestive secretion, Trappmann says, is produced in the plasma of the cells in the form of very fine granulations, which soon increase in size and numbers and assemble

187

mostly in the apical parts of the cells. Then, either the granule-containing cells are discharged from the epithelium or the apices of the cells are cut off and liberated with their contents. The secretion is finally set free by a dissolution of the cell walls.

The destruction of the epithelial cells in the process of secretion is compensated by the addition of new cells formed in the regenera-

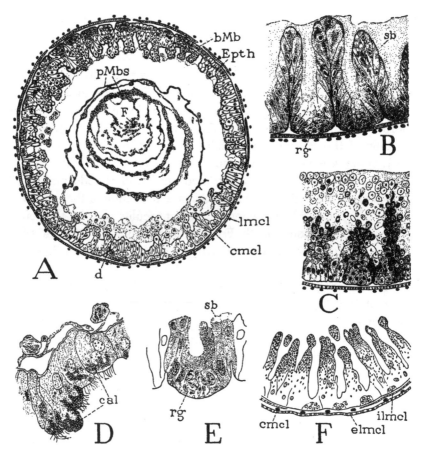

Fig. 73. Histological details of the ventriculus of a worker.

A, cross section of ventriculus and peritrophic membranes. B, epithelium with cells in resting condition. C, epithelium proliferating small digestive cells into covering layer. D, epithelium containing calcium particles (from Fyg, 1932). E, longitudinal section of fold of epithelium with regenerative crypt, showing one cell in mitotic division (from J. and C. Evenius, 1925). F, section of epithelium containing spores of *Nosema apis* (from White, 1918).

d, attachment of peritrophic membrane to ventricular epithelium. For explanation of abbreviations see pages 199–200.

tion centers (fig. 73 B, *rg*). These regeneration "crypts" are groups of small cells of the epithelium at the bottoms of the epithelial pits. By mitotic division (E) the crypt cells proliferate new cells inward that replace those depleted by secretion activities. According to J. and Christa Evenius (1925), however, mitoses in the regeneration cells are not often observed, so that renewal of the epithelium evidently takes place gradually. Yet, since division is never seen in the functional epithelial cells, regeneration must proceed only from the crypt cells.

In addition to its secretory function, the ventricular epithelium serves also for excretion, particularly of calcium. The presence of granules of calcium salt in the epithelial cells of the bee was first observed by Koehler (1920), and the granules were later shown by Fyg (1932) to be calcium carbonate. The calcium appears in the form of minute crystals massed in the inner ends of the ventricular cells (fig. 73 D, *cal*) and occurs also in the Malpighian tubules. Calcium is commonly thus excreted by insects; its excretion, as noted by Koehler, is probably merely a process of getting rid of the calcium unavoidably taken in with the food, the insect having little need of calcium in its tissues. In cases of infestation by *Nosema apis* the spores of the organism accumulate in great numbers in the walls of the ventriculus (F). According to Hassanein (1953), the secretory cells of the ventricular epithelium become reduced to a disorganized meshwork of broken-down cells filled with spores. The spores are found also in the Malpighian tubules and have a deleterious effect on the rectum, which frequently becomes greatly distended with a watery fluid leading to dysentery.

The Peritrophic Membrane— Transverse sections of the ventriculus of an adult honey bee (fig. 73 A) show that the food mass or remnant of it (F) in the lumen is enclosed in several thin, irregularly concentric coverings. These cylindrical food envelopes are the *peritrophic membranes* (*pMbs*). Whitcomb and Wilson (1929) say there may be as many as 12 membranes present at one time in the bee, and Weil (1935) reports finding 20 of them in a 10-day-old bee; but ordinarily there are fewer than either of these records, and only one membrane may be present. Since the peritrophic membranes surround the food, they are permeable in one direction to the digestive secretions and in the other to the digested food (see review by Day and Waterhouse, 1953). To the question as to why

189

insects have a peritrophic membrane there is no positive answer. It has been suggested that the membrane serves to protect the ventricular epithelium from abrasion by coarse foods, such as spiny pollen grains in the stomach of the honey bee, but some liquid-feeding insects also have peritrophic membranes, and some that eat solid food have none.

The chemical nature of the peritrophic membrane, whether chitinous or not, is a question on which investigators have not agreed. The peritrophic membrane of the honey bee was reported by Campbell (1929) and by von Dehn (1933) to be chitinous; on the other hand, Pavlovsky and Zarin (1922), Weil (1935), Hering (1939), and Kusmenko (1940) say it is nonchitinous. Weil suggests that those who reported the presence of chitin probably got the intima of the stomodaeal valve mixed with their test material. Waterhouse (1953), however, obtained positive reactions for chitin in tests on the peritrophic membranes of a large number of other insects, and he says that from the facts now available there is overwhelming evidence that the food-enclosing membranes in many insects contain chitin. The principal objection to admitting that the peritrophic membranes are chitinous comes from the deep-seated idea that the endoderm cannot be a chitin-forming layer. Yet both the endoderm and the ectoderm originally are parts of the blastoderm.

The origin and manner of formation of the peritrophic membrane are also subjects on which there has been much difference of opinion. In some insects, particularly in the higher Diptera, the membrane appears to be formed entirely by secretion from a ring of ventricular cells around the base of the stomodaeal valve; in most other insects it is produced from the entire length of the ventricular epithelium. As we have seen, both types of membrane, distinguished as type I and type II by Waterhouse (1953), are present in the honey-bee larva (fig. 68 C); in the adult bee only type II is formed. Butt (1934) contends that the cells of the ring around the base of the stomodaeal valve that secrete a type-I membrane are derived from the stomodaeum, but in the honey-bee larva they very clearly belong to the ventriculus (C, rnCls).

In the adult honey bee there can be no doubt that the peritrophic membranes are formed consecutively from the entire length of the ventricular epithelium. In the section shown at A of figure 73 a part of the outermost membrane is still in continuity with the gelatinous covering of the epithelial cells. This layer evidently represents the thick type-II membrane of the larva, but in the adult it has an apparently denser inner surface, the *Grenzmembran* of Trappmann (1923), which in sections appears as a dark border. During secretion activity the gelatinous layer becomes filled with the bodies discharged from the cells containing the digestive enzymes (C). Then it appears that at least the inner part of this mass is separated from the epithelium but that only its thickened surface layer becomes a peritrophic membrane (A), the rest with the inclusions presumably being dissolved to liberate the digestive elements. Pavlovsky and Zarin (1922) describe the peritrophic membrane as a thick secretion from the epithelial cells containing in its depth, as in a sponge, the digestive secretions of the stomach, "on account of which the same quantity of ferment is capable of acting for a longer period upon the food."

Trappmann (1923) describes the peritrophic membrane of the bee as a transformation of the striated border, or rhabdorium, of the ventricular epithelium produced by the secretory activity of the cells. In the resting condition the rhabdorium is covered by a *Grenzmembran*. The discharge of secretion in quantity into the rhabdorium, however, separates this border membrane from the epithelium, and it then breaks away as a peritrophic membrane. A repetition of the process forms in succession a number of peritrophic membranes, which become tubes enclosed one within the other. Whitcomb and Wilson (1929) give about the same explanation of the origins of the peritrophic membranes from the rhabdorium, but they observe that the striated border is given off with each membrane and that a new border is formed on the epithelial cells. Weil (1935) also describes the peritrophic membranes of bees and wasps as arising from the rhabdorium. The latter, he says, is a plasmatic formation of the epithelial cells, which toward the lumen may be thickened to a dense gelatinous layer having a sharp margin, the *Grenzmembran* of Trappmann. From its intense blue coloring in Azan preparations, Weil concludes that this inner layer, the *Hüllschicht*, is strongly impregnated with digestive enzymes. The pressure of the extruded secretion separates this layer as a peritrophic membrane, and the rhabdorium is broken, part of it adhering to the border membrane and part to the epithelium.

Von Dehn (1933) describes the peritrophic membrane as a direct

191

product of the epithelial cells. She says it is formed as a fine chitinous layer given off from the surfaces of the cells at the *bottom* of the striated border, or *Stäbchensaum*, along the whole length of the ventriculus. Then, by secretion from the cells, the membrane is pushed through the striated border and comes to lie on the surface of the latter, where it appears as the *Grenzmembran* of Trappmann. Finally it is separated and becomes a free peritrophic membrane in the ventricular lumen. In this manner, von Dehn says, the membrane is formed not only in Hymenoptera but in insects of other orders, and she illustrates her claim convincingly from sections of the ventriculus of *Vespa* showing the membrane in stages of its progress through the striated border. If formed in this way as a direct product of the cells, it is more understandable that only the membrane itself should be chitinous.

The Proctodaeum— The ventriculus of the adult bee opens directly into the proctodaeum (fig. 76 F). The epithelium at its posterior end may project somewhat as a fold over the mouth of the intestine, but it does not extend into the latter to form a ventricular valve, except at the first opening of the ventriculus into the proctodaeum in a late larval stage (C, *VVlv*).

The proctodaeum is differentiated into two principal regions, which are the *anterior intestine* (fig. 70, *AInt*) and the *posterior intestine*, or *rectum* (*Rect*). The anterior intestine is a relatively slender tube looped upon itself in going from the posterior end of the ventriculus in the anterior part of the abdomen to the posterior rectum. In the bee the anterior intestine is appropriately termed the *small intestine*, or *Dünndarm*, but in some insects it is as large as or even larger than the rectum. Its epithelial wall is thrown into six longitudinal folds (fig. 74 A) that continue through most of its length. On the outside is a thick layer of circular muscle fibers, but this part of the intestine has no longitudinal muscles. At its anterior end the intestine is somewhat widened where it joins the ventriculus, and this part is known as the *pylorus*. The Malpighian tubules open into the pyloric lumen immediately behind the ventriculus (fig. 76 F, *Mal*). Just posterior to their mouths is a thick fold of the intestinal wall, forming a *pyloric sphincter*, or *valve* (*pyVlv*), behind which the intestinal intima is minutely spinous. The posterior end of the anterior intestine tapers to its union with the rectum. The six folds of its inner wall here become crowded and form a puckered exit

192

orifice, which in some specimens may be pushed into the rectum as a small circular fold (fig. 74 B).

The rectum is a large, thin-walled sac lying in the posterior part of the abdomen (fig. 70, *Rect*). Anteriorly it is abruptly narrowed where it is joined by the slender anterior intestine, and posteriorly

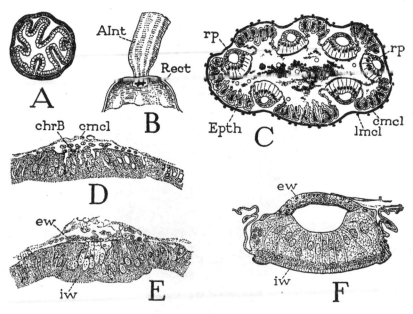

Fig. 74. Structural details of the proctodaeum and the rectal pads.

A, cross section of anterior intestine. B, opening of anterior intestine into rectum. C, cross section of rectum through rectal pads; D, section of rectal pad region of rectal wall of pupa, showing discharge of chromatic bodies (from Dobrovsky, 1951). E, same of older pupa showing formation of outer wall of rectal pad (from Dobrovsky, 1951). F, section of mature rectal pad (from Trappmann, 1923).

For explanation of abbreviations see pages 199–200.

it contracts to a tapering tube that opens at the anus on the much-reduced tenth abdominal segment. The rectal epithelium for the most part is a thin, nucleated protoplasmic layer lined with a thin cuticular intima and is thrown into numerous small folds (fig. 74 C). On the outer surface is a sheath of closely placed circular muscle fibers (*cmcl*) and an external layer of smaller, widely separated longitudinal fibers (*lmcl*). The folds of the epithelium and intima give the rectal sac a great distensibility and allow it to hold an immense accumulation of fecal matter and excretion from the Malpighian tubules. Bees never eject their feces in the hive, and during prolonged cold periods in winter they are forced to retain so much intestinal waste that the rectum becomes expanded to an enormous bag occupying all available space in the abdomen (fig. 75).

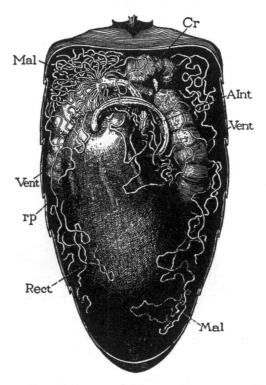

Fig. 75. Interior of abdomen of over-wintered worker before first flight, showing the rectum distended with undischarged feces.

For explanation of abbreviations see pages 199–200.

In the anterior part of the rectum are six long, regularly spaced thickenings of the epithelial wall, which are characteristic features of the rectum in most adult insects and formerly were known as "rectal glands." Since no secretory activity has ever been observed in these organs, they are now termed *rectal pads* or *rectal papillae* according to their form. When the rectum of the bee is distended, the pads appear on the outer surface as six opaque ridges (fig. 70, *rp*). In section, each pad is seen to be a hollow tube with a thick inner wall (fig. 74 F, *iw*) composed of a single layer of very large cells and a thinner external wall (*ew*) formed of two layers of small cells. The inner walls of the pads are continuous with the thin folded epithelium of the rectal wall between the pads. Toward the lumen of the rectum the pads present smooth, evenly convex surfaces, the cuticular intima of which is thickened along the edges of the pads, forming a prominent shoulder on each side (F).

In most insects the rectal pads are simple thickenings of the rectal epithelium; if they have a lumen it is an opening from the body cavity. The inner walls alone of the pads in the bee, therefore, evidently represent the usual single-layered organs, and the outer walls appear to be something secondarily added. It was claimed by C. Evenius (1933) and by Lotmar (1945) that the cells of the outer walls of the pads are adventitious mesoderm cells from the body cavity that are distributed around the rectum and gradually become layered on its wall. The outermost of these cells, according to Evenius, form the rectal muscles, the others form the two-layered outer walls of the pads. Dobrovsky (1951), however, offers an entirely different explanation of the origin of the outer-wall cells. During the early part of the pupal period, he says, there takes place in the anterior part of the rectum a discharge of large numbers of chromatic bodies from the epithelial cells in the region of the pads (fig. 74 D, *chrB*). These bodies are retained within the muscle sheath (*cmcl*) and here become small cells that finally assemble to form the outer layers of the rectal pads (E, *ew*). According to Evenius and Lotmar, the pads are ectodermal and mesodermal structures; according to Dobrovsky, they are entirely ectodermal.

The function of the rectal pads is more obscure than their structure. The organs are evidently merely specialized parts of the six epithelial folds that usually run through the intestine, but they differ much in size, form, and structure in different insects, which fact is difficult to understand if they all have the same function. Since there is no evidence that the organs in any case are glands, Wigglesworth (1932) has advanced the idea that they serve for the absorption of excess water from the rectal contents, the conservation of water being an important matter with terrestrial insects. He gives abundant evidence that water is absorbed from the rectum, and he suggests, therefore, that the most probable site of absorption is the rectal pads. Yet it might be questioned whether or not the structure of the organs in all cases is adapted to absorption. In the bee the organs have a thin but apparently dense intima covering their convex inner surfaces, with thickened margins on the sides (fig. 74 F), so that they do not appear to be constructed specifically for an absorptive function. It might be supposed, however, that when the rectum is contracted the pads exert a pressure on the rectal contents that facilitates the extraction of water, which then is absorbed through the thin parts of the rectal walls. Any explanation leaves unexplained the cavities in the pads of the bee.

THE MALPIGHIAN TUBULES

The Malpighian tubules are appurtenances of the alimentary canal, but they are excretory organs and therefore have an independent functional status. The tubules of the adult honey bee are long, whitish, convoluted, threadlike tubes, perhaps a hundred of them (fig. 70, *Mal*), wrapped and coiled around one another and about the viscera in the abdominal cavity, all opening separately into the anterior end of the intestine. The wall of a tubule is formed of a single layer of cells (fig. 76 E) with an inner striated border (*sb*) and an outer basement membrane. According to Trappmann (1923), there is an enveloping peritoneal sheath of connective tissue, in which are flat bands of striated muscle fibers that surround the tubule in wide spirals. Morison (1928a) says the muscles vary from one to four on each tubule, "around which they are twisted in a spiral which encircles the tube about three times." The muscle fibers are accompanied by long, wavy tracheoles.

The excretory products of the Malpighian tubules of the bee have not been analyzed, but in other insects they appear in the form of small crystals of nitrogenous and other substances, the most usual being urates, leucin, phosphates, calcium oxalate, and calcium carbonate. The excretory matter in the honey bee, as shown by Trappmann (1923), first accumulates as a secretion in the inner ends of the tubule cells (fig. 76 G) and is then extruded through the striated border in free globules constricted off from the cells (H). If the discharged globules are small, the striated border simply closes again, but after abundant excretion it is re-formed. Expulsion of the matter is effected by contraction of the tubule muscles; in a living condition the movements of the tubules may be observed.

In the larva of the honey bee there are only four Malpighian tubules, which in the mature larva are long, thick tubes extending forward along the ventriculus as far as the thorax (fig. 68 A, *Mal*). During the feeding stage of the larva the tubules are not open into the intestine, but their narrowed proximal ends are inserted between

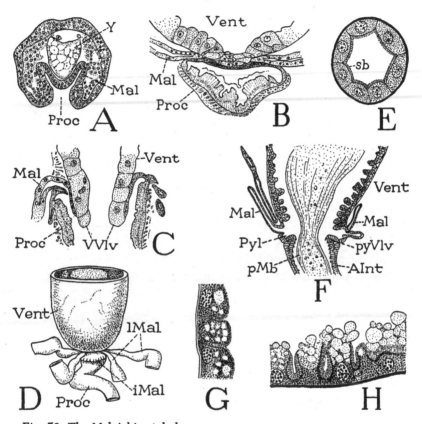

Fig. 76. The Malpighian tubules.

A, section of posterior end of 52–54-hour embryo, Malpighian tubules carried in on end of proctodaeum (from Nelson, 1915). B, blind ends of Malpighian tubules in septum between ventriculus and proctodaeum of larva just before sealing (from Nelson, 1924). C, opening of larval tubules and ventriculus into proctodaeum in larva 12 hours after sealing (from Nelson, 1924). D, junction of ventriculus and proctodaeum in mature larva, showing bases of larval tubules and rudiments of imaginal tubules (from Dodbrovsky, 1951). E, cross section of tubule in nonsecretory condition (from Trappmann, 1923). F, section of alimentary canal of adult bee at junction of ventriculus and proctodaeum, showing openings of Malpighian tubules into pyloric region (from Armbruster, 1931). G, section of wall of Malpighian tubule in beginning of secretion (from Trappmann, 1923). H, Malpighian epithelium in active secretion (from Trappmann, 1923).

For explanation of abbreviations see pages 199–200.

the ventriculus and the proctodaeum (fig. 76 B, *Mal*). After the larva has been sealed in its cell, however, and the ventriculus has opened into the intestine, the Malpighian tubules break through the proctodaeal layer of the septum (C) and discharges their accumulated contents into the intestine along with the refuse from the ventriculus. In the pupal stage the larval tubules completely degenerate and disappear.

The tubules of the adult honey bee are formed as new organs quite independent of the larval tubules. While the larval organs are still present (fig. 76 D, *lMal*), the imaginal tubules appear as a ring of numerous small buds (*iMal*) on the anterior end of the intestine immediately behind the bases of the larval tubules. According to Oertel (1930), the imaginal buds appear about five hours before the sealing of the larval cell and grow during the pupal stage by cell proliferation. Lotmar (1945) says that in a 7-day-old pupa the epithelium of the imaginal tubules has a typical striated border and already shows definite evidence of secretion. In most other holometabolous insects the tubules of the adult are reconstructed from the larval tubules.

The origin of the Malpighian tubules is a subject on which there has been much difference of opinion. Generally the tubules are regarded as outgrowths of the pyloric region of the proctodaeum, and therefore as derivatives of the ectoderm, but some investigators strongly insist that they open into the mesenteron and are therefore endodermal. A clear case of the tubules opening into the mesenteron appears to be present among the Ephemeroptera in the family Baetidae, in species of which, as shown by Tirelli (1929) and Grandi (1950), the tubules open some distance before the union of the mesenteron with the proctodaeum. Trappmann (1923) and Weil (1935) both assert that the tubules of the adult honey bee open into the mesenteron. On the other hand, from the observations of Nelson (1915) and Kusmenko (1941) there seems to be little doubt that the larval tubules of the honey bee are ectodermal. According to Nelson, the rudiments of the larval organs appear in a 44–46-hour embryo as four pitlike depressions of the ectoderm at the posterior end of the body around the point where the proctodaeum will be formed. Kusmenko confirms Nelson's statement that the tubules of the bee have their inception on the outer surface of the embryo, and both writers cite references in which the same

198

thing has been recorded for other insects, though in most such cases the tubules appear around the already-formed proctodaeum. In the honey bee, as the proctodaeum grows inward (fig. 76 A, *Proc*), it carries the Malpighian tubules on its anterior end.

Since the larval tubules of the honey bee thus appear unquestionably to be ectodermal in their origin and eventually discharge into the proctodaeum, the imaginal tubules that grow out *behind* their bases (fig. 76 D) must also belong to the ectodermal proctodaeum. In the adult bee the tubules open at the apparent junction of the ventriculus with the intestine, but as seen at F of figure 76, taken from Armbruster (1931), it is evident that the tubules open into a short pyloric region (*Pyl*) of the proctodaeum. Structurally and in their functional activity, it must be admitted, the Malpighian tubules resemble the ventriculus. They appear to lack a chitinous intima, the epithelial cells have a striated border, and the secretory processes are similar to those of the ventriculus. These features common to the Malpighian tubules and the ventriculus, therefore, are evidently to be interpreted as common adaptations to function, rather than as evidence of origin from the same germ layer.

Explanation of Abbreviations on Figures 64–76

AInt, anterior intestine.
AMR, anterior mesenteron rudiment.
An, primary anus.
An', definitive anus.

Blc, blastocoele.
bMb, basement membrane.
Bpr, blastopore.
Br, brain.

cal, calcium particles.
Cb, cibarium.
ChrB, chromatic bodies.
cmcl, circular muscles.
Cr, crop ("honey stomach").

Ecd, ectoderm.
elmcl, external longitudinal muscles.
End, endoderm.

Epth, epithelium.
ew, external wall of rectal pad.

FGld, food gland.

GB, germ band.
Gcl, gastrocoele.

HGld, head salivary gland.
Hphy, hypopharynx.
Ht, heart.

ilmcl, inner longitudinal muscles.
iMal, imaginal Malpighian tubules.
iw, inner wall of rectal pad.

Lb, labium.
Lm, labrum.
lMal, larval Malpighian tubules.
lmcl, longitudinal muscles.
LP, lateral plate of germ band.

199

Mal, Malpighian tubules.
mcls, muscles.
Ment, mesenteron.
MP, middle plate of germ band.
Msd, mesoderm.
Mth, primary mouth.
Mth', definitive mouth.

Oe, oesophagus.

Phy, pharynx.
pMb, peritrophic membrane.
pMbs, peritrophic membranes.
PMR, posterior mesenteron rudiment.
Proc, proctodaeum.
Prvent, proventriculus.
Pyl, pylorus.
pyVlv, pyloric valve.

Rect, rectum.
rg, regenerative cells.
rncls, ring cells.
rp, rectal pad.

sb, striated border.
skGld, silk gland.
slGld, salivary gland.
SoeGng, suboesophageal ganglion.
Stom, stomodaeum.
SVlv, stomodaeal valve.

ThGld, thoracic salivary gland.

Vag, vagina.
Vent, ventriculus (mesenteron of embryo).
VNC, ventral nerve cord.
VVlv, ventricular valve.

Y, yolk.

X

THE CIRCULATORY SYSTEM

A COMPLEX animal could not exist if its body were made up entirely of solid organs and fixed tissues, any more than a city would be habitable if built of masses of houses continuous in all directions. In each there must be spaces given over to open highways along which vehicles or a circulating medium can distribute the necessities of life from receiving stations to consumers. In the animal the liquid blood, flowing through tubes or through spaces between the organs, plays the role of distributor, carrying the nutritive substances from the alimentary canal to the cells of the tissues where they are consumed. The blood carries all the food elements received from the alimentary canal to all the cells of the body; the cells take out what they individually need for their own particular functions. The blood of vertebrate animals carries also ogygen from the gills or lungs and removes carbon dioxide, but most insects have separate air tubes for the delivery of oxygen. Finally, the blood acts as ash collector by taking the waste products of metabolism from the tissues and carrying them to special organs of excretion. In most animals the blood is kept in motion by a pumping apparatus, which, together with the blood itself and the blood vessels or other channels through which the blood flows, constitutes the *circulatory system.*

The circulatory system of an insect includes almost the entire body, since the blood fills all the space in the body cavity not occupied by other tissues. The blood is kept in motion principally by a median, tubular, pulsating *dorsal vessel* extending from near the posterior end of the abdomen into the head beneath the brain. The blood enters the abdominal part of the vessel through paired lateral *ostia* and is driven forward by progressive contractions of the mus-

201

cular walls of the vessel into the head, where it is discharged from the open anterior end of the vessel. The part of the dorsal vessel containing the ostia is commonly distinguished as the *heart*, and the narrower anterior imperforate part as the *aorta*. From the head the blood flows backward through more or less definite channels among the organs and fat cells of the body, and also through the legs and the wings, finally to re-enter the ostia of the heart, but it may be assisted in its circuit by various accessory pulsating organs. Chief of the latter are a muscular *dorsal diaphragm* stretched across the upper part of the abdominal cavity beneath the heart and a muscular *ventral diaphragm* over the nerve cord in the abdomen. The dorsal diaphragm in some insects vibrates rhythmically in an anterior direction; the ventral diaphragm vibrates backward. The cavity above the dorsal diaphragm is the *dorsal sinus*, or *pericardial cavity;* that below the ventral diaphragm is the *ventral sinus*, or *perineural cavity*. The diaphragms have free edges between their points of attachment on the body wall so that the blood has access to the sinuses. In addition to the diaphragms there may be in various parts of the body, especially in the head and thorax, accessory pulsating organs in the form of vesicles or membranes.

Most arthropods other than insects have a variously developed system of arteries branching from the dorsal vessel, but in all of them the blood in some part of its course flows through open channels of the body cavity before it re-enters the heart. Among the annulate animals a completely closed circulatory system occurs only in some of the annelid worms; in the arthropods it appears that there has been a progressive breakdown of the closed system. Only a few insects, notably the cockroach and the mantis (see McIndoo, 1939, Nutting, 1951), retain remnants of lateral arteries from the dorsal vessel.

DEVELOPMENT AND EVOLUTION OF THE CIRCULATORY ORGANS

The circulatory system is so closely tied up with the development of the mesoderm that we can best understand it in the insects by following the history of the mesoderm from the annelids into the arthropods as it may be interpreted from ontogeny. The primary body cavity containing the body liquid, or blood, is the blastocoele of the early embryo (fig. 65 A, *Bcl*). The blastocoele is first invaded by the endoderm (B, *End*) and then by the mesoderm (C, *Msd*), but since it still contains the blood it is now known as the *hemocoele* (fig. 77 A, *Hcl*). The mesoderm of the annelid worm originates at the posterior end of the embryo, but it grows forward in the hemocoele as two lateral bands (*Msd*) extending forward to the mouth.

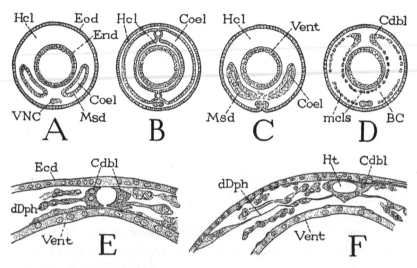

Fig. 77. Evolution and embryonic development of the heart and the body cavity.
A, diagram of an early developmental stage, hemocoele invaded by mesodermal coelomic sacs. B, diagrammatic cross section of an annelid worm, hemocoele almost replaced by a coelome. C, diagrammatic early stage of insect embryo with mesodermal coelomic pockets. D, later stage of same, coelomic walls disrupted, uniting hemocoele and coelome in a definitive body (*BC*). E, cross section of dorsal part of embryo of honey bee, cardioblasts not yet united (from Nelson, 1915). F, later stage of same, cardioblasts united to form the heart (from Nelson, 1915).
For explanation of abbreviations see page 216.

Each band then becomes hollowed by a series of segmental cavities (*Coel*), so that the mesoderm is thus divided into paired pouches corresponding with the body segments of the worm. The mesodermal cavities will form a new body cavity termed the *coelome*, and the segmented pouches are therefore known as the *coelomic sacs*. The two sacs of each segment expand around the alimentary canal (B) until they meet above and below it, with the result that the coelome (*Coel*) all but supplants the original hemocoele (*Hcl*), which is reduced to a narrow peripheral space against the ectoderm and a median space around the alimentary canal. The walls of the mesoderm become muscles and other tissues, but the inner surface becomes an epithelium lining the coelomic sacs, which in the worm entirely shuts off the coelome from the hemocoele. The blood confined to the reduced hemocoele is eventually enclosed in blood vessels, including a longitudinal dorsal and a ventral vessel with interconnecting lateral branches.

In the insect embryo the mesoderm bands (fig. 77 C, *Msd*) likewise grow upward in the sides of the hemocoele and become excavated by coelomic cavities, but in most insects the coelome is not divided by partitions into segmental compartments. In the honey bee, as shown by Nelson (1915), the coelomic cavities are only very small cleavage spaces formed in the lateral or upper parts of the mesoderm bands (C, *Coel*). Instead, now, of the coelomic sacs expanding until they almost obliterate the hemocoele, as they do in the worms, most of their walls in the insect break up into cells that become muscles and fat tissue (D), and they do not form a peritoneum. Consequently the coelomic cavities and the hemocoele become completely merged in the definitive body cavity (*BC*), which is thus really a *mixocoele*, though it is called the hemocoele because it is mostly of hemocoelic origin and contains the blood. The lateral, or finally the uppermost, cells of the mesoderm, however, do not break up. These cells (D, *Cdbl*) are destined to form the dorsal blood vessel, or heart, and are hence known as the *cardioblasts*. The opposing bands of cardioblasts approach each other along the middorsal line of the hemocoele and become crescentic in section with their concave faces toward each other (E, *Cdbl*). Finally their edges unite, forming a mesodermal tube which becomes the dorsal blood vessel (F, *Ht*). It thus comes about that the blood of the insect occupies the body cavity, except that which is temporarily contained in the dorsal vessel.

The dorsal diaphragm is formed from strands of mesodermal cells that extend laterally from the cardioblasts (fig. 77 E, *dDph*), which eventually form sheets of tissue that connect the heart with the ectoderm of the dorsal body wall (F, *dDph*). The ventral diaphragm, Nelson (1915) says, is formed in the bee "from muscle fibers arising near the ventral longitudinal muscles, which extend out toward the mid-line to join those of the opposite side."

THE LARVAL CIRCULATORY ORGANS

The heart of the honey-bee larva is described by Nelson (1924) as a thin-walled tube, widest at its blind posterior end in the ninth abdominal segment and tapering anteriorly to the front of the mesothorax. Here it turns downward beneath the anterior tracheal loop and decreases rapidly in diameter, being continued as the aorta into the head. The heart walls formed of the cardioblasts contain bundles of delicate striated muscle fibrils. Externally the larval heart is clothed by a loose membrane of minute, branched connective tissue cells, processes from which serve to anchor the heart to the dorsal ectoderm. In the mesothorax, the metathorax, and each of the first nine abdominal segments the heart is constricted and perforated by a pair of ostia, the lips of which form valvelike flaps projecting inward and forward into the heart lumen. The aorta is open on its ventral side, having in transverse section the form of an inverted U; its anterior part goes beneath the brain and ends at the anterior face of the latter.

The dorsal diaphragm of the larva, according to Nelson, is well developed only from the fourth to the ninth abdominal segment, inclusive, and in structure is similar to that of the adult (fig. 81). Its lateral margins are attached to the body wall at points corresponding to the intersegmental lines. The diaphragm consists of two very delicate noncellular membranes attached medially to the ventral wall of the heart. Between the membranes are irregular rows or strings of diaphragm cells disposed in a lacework pattern. An integral part of the diaphragm consists of fan-shaped groups of muscle fibers arising laterally on the body wall at the points of attachment of the diaphragm and spreading mesally to the heart. Anterior to the fourth abdominal segment the diaphragm becomes so delicate that its structure is difficult to follow.

The ventral diaphragm, Nelson says, is well developed in a newly hatched larva, in which it is a continuous sheet of transverse muscle fibers over the nerve cord. In older larvae, however, it becomes reduced to delicate, more or less isolated fibers forming only a loose and insignificant meshwork.

205

THE CIRCULATORY ORGANS OF THE ADULT BEE

The circulatory organs of the adult honey bee have been elaborately described by Freudenstein (1928). They include the dorsal vessel, a dorsal diaphragm, a ventral diaphragm, and pulsating organs in the head and thorax.

The Dorsal Vessel— In the adult bee the dorsal vessel (fig. 78, *Ao, Ht*) extends from the posterior half of abdominal segment VI

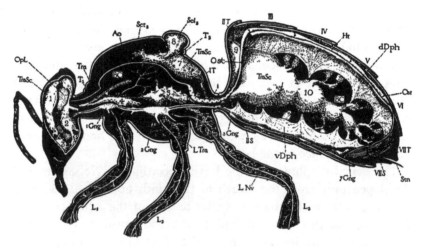

Fig. 78. Body of worker bee cut longitudinally, muscles and alimentary canal removed, exposing dorsal blood vessel, diaphragms, tracheae, air sacs, and ventral nerve cord.

i, convoluted part of aorta. For explanation of abbreviations see page 216.

into the head. It has thus been shortened by the length of three segments since the larval stage when it reached back into the ninth segment. The heart, or part of the tube lying in the abdomen (*Ht*), increases in width to the posterior end (fig. 79 B) and is perforated by five pairs of lateral ostia (*Ost*), the first in segment II of the abdomen (fig. 78), the last in segment VI. In segment III the heart turns downward over the anterior edge of the dorsal diaphragm (*dDph*) and becomes a narrow tube continued as the aorta beyond the first ostia. Where the aorta enters the thorax through the petiole, it is thrown into a series of loops (*i*) enclosed in a delicate transparent sheath. Then again becoming a narrow tube, the aorta (*Ao*) arches upward and forward between the dorsal longitudinal muscles of the thorax and finally goes through the neck into the head, where it ends openly beneath the brain.

The heart is a thick-walled muscular tube composed of semicircular muscle fibers joined dorsally and ventrally (fig. 82 A, *Ht*), giving evidence of the origin of the heart from paired rows of cardioblasts (fig. 77 E, F), the union of which is still marked on the surface by pale median dorsal and ventral lines (fig. 82 B). The muscular structure of the heart becomes weaker anteriorly but is continued as far as the convoluted part of the aorta, beyond which it is extremely faint and gives way to flat nucleated cells. The heart walls have no endothelial lining other than the sarcolemma of the muscles. The heart is suspended from the dorsal body wall by fine connective tissue strands (fig. 82 A), and fibrils from the diaphragm and its muscles are attached on its lateral and ventral walls.

The ostia are somewhat obliquely vertical slits in the lateral walls of the heart (fig. 82 B, *Ost*), which open into the lumen between long flaps extending inward and anteriorly. When the heart contracts (fig. 79 C), the flaps permit the blood to be driven forward but prevent its backward flow. During dilatation of the heart the ostial lips open (D) and allow free entrance to the blood. The ostial lips are the only valvular structures within the heart and divide the heart into so-called chambers.

The heart is abundantly supplied with tracheae given off from segmental branches of the great lateral air sacs of the abdomen (fig. 78). The branches themselves divide and swell into air sacs in the dorsal sinus, from which arise numerous small branching tracheae that go directly to the heart walls (fig. 80).

In action the heart ordinarily beats from behind forward by successive waves of contraction running anteriorly. The systolic phase is produced by the heart muscles. It was formerly supposed that the diaphragm muscles are responsible for the dilatation of the heart, but the heart will continue its rhythmic pulsations when the diaphragm is cut loose from the body wall. In various insects the heart periodically reverses the direction of its beat, but reversal has not been observed in bees or ants (Gerould, 1933).

The rhythmic beating of an insect's heart is readily observed, but the source of its stimulation is not so easily determined. There has been much discussion as to whether the heart beat is generated by nerve stimulus (neurogenic action) or arises in the heart muscles

207

themselves (myogenic action). Lateral heart nerves originating anteriorly from the stomodaeal system have been described in some insects; in others the heart is said to be innervated from branches of the ventral nerve ganglia, or from both sources. Morison (1928a)

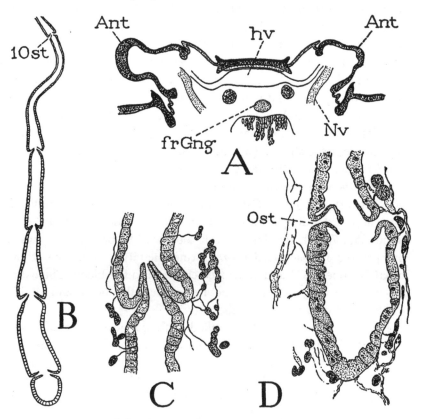

Fig. 79. Structural details of the heart and the pulsatile vesicle of the head (from Freudenstein, 1928).

A, horizontal section through anterior part of head, antennal bases, and head vesicle. B, diagrammatic horizontal section of heart, showing ostia. C, section of heart through third pair of ostial valves, heart in contraction. D, section of last heart chamber in condition of expansion.

For explanation of abbreviations see page 216.

reports that in about 60 hearts of the three castes of the honey bee examined by various staining methods he failed to find any trace of heart nerves. Rehm (1939) from experiments in cutting the heart of the bee free from other connected tissues, and even in cutting the heart itself into segmental sections, found that in all cases the heart continued to beat. His results thus seemed to show that the heart beat does not depend on nerve connections from any source. However, in one case he was able to trace a fine branch from a nerve of the diaphragm muscles, originating in the ventral nerve cord, that ended against the heart wall in a small group of ganglionic nerve cells. These cells, therefore, Rehm concluded, may be an automatic center for the heart stimulation, even when the nerve is severed. If so, it is to be assumed that similar groups of cells remain to be discovered on the other heart segments. Fortunately for the bee, in spite of our ignorance the heart beats.

The Dorsal Diaphragm— The dorsal diaphragm stretches across the upper part of the abdominal cavity (fig. 78, *dDph*) from the anterior part of segment III, beginning here with a free margin beneath the heart, and continues posteriorly into segment VI. The margins of the diaphragm are drawn out into points attached to the terga of segments III to VI inclusive at the bases of the antecostal apodemes (figs. 80, 81, *a*). The free borders between the attach-

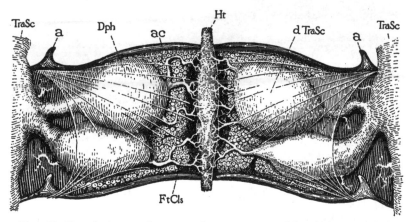

Fig. 80. Dorsal sinus and contained organs in an abdominal segment, seen from below through the transparent diaphragm. *a*, antecostal apodeme of tergum. For explanation of abbreviations see page 216.

ments form deep semicircular notches. Though the diaphragm is very thin and is transparent in the natural condition, it is a complex structure, formed apparently of two noncellular limiting membranes between which are muscle fibers and a layer of scattered cells. The heart (*Ht*) is seen traversing the middle of the diaphragm, to which it is firmly attached.

The diaphragm muscles consist of five paired, fan-shaped groups of fibers (fig. 81, *DphMcl*) radiating toward the heart from the lateral points of attachment of the diaphragm on the tergal plates. The fibers are very slender, being from 10 to 20 microns in width, and many of them are branched. They have a distinct cross striation, and in appearance they resemble the fibers of other body muscles. According to Morison (1928a), the nuclei lie in the axes of the fibers, but here and there a nucleus may be seen lying apparently in an outer layer of sarcoplasm. Some of the anterior and posterior fibers of each group end in the diaphragm (fig. 81), but most of

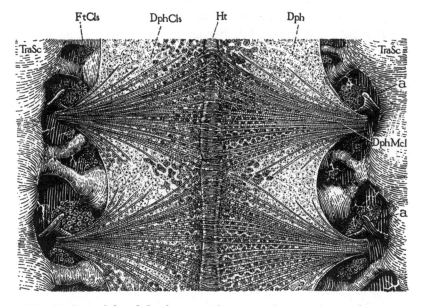

Fig. 81. Part of dorsal diaphragm with associated organs from a drone, seen from below.

a, antecostal apodeme of tergum. For explanation of abbreviations see page 216.

them reach to the edge of the heart. Here the fibers abruptly end (fig. 82 C) and give off spreading brushes of fine branching threads, some of which are attached on the wall of the heart, while others join beneath the heart with threads from the opposite side in an intricate network.

The diaphragm cells are loosely scattered toward the edges of the diaphragm (fig. 81, *DphCls*), but for the most part they form everywhere between the muscle fibers irregular groups and long branching and reuniting bands in which cell boundaries are often indistinct (fig. 82 D). In section, the rows of cells (A) are seen to have definite upper and lower borders as if enclosed between two membranes, but the intervening spaces are bridged by an apparently single membrane attached along the edges of the cells. Nelson (1924) suggests that the diaphragm membranes may be of the nature of basement membranes of the cells themselves. The diaphragm cells are commonly called the "pericardial cells," which term is appropriate if the diaphragm is regarded as a pericardium, but it does not actually enclose the heart. The dorsal diaphragm has a pulsating movement independent of that of the heart, in which waves of contraction run forward.

Lying within the diaphragm close along the sides of the heart are masses of cells quite different from the other diaphragm cells. These are relatively large, oval, disconnected cells, most of which are distinctly binucleate; they are here distinguished as *paracardial cells* (fig. 82 C, *PaCls*). They lie over the diaphragm muscles and the other diaphragm cells, but as seen in section (A) they are covered by a dorsal membrane of the diaphragm and are interlaced by branching strands from the wall of the heart. Freudenstein (1928) and other writers apparently have not distinguished these paracardial cells from the ordinary diaphragm cells.

The Dorsal Sinus— The dorsal sinus is merely the part of the body cavity above the dorsal diaphragm; it is in free communication with the visceral cavity of the abdomen through the openings along the edges of the diaphragm. The sinus contains the dorsal tracheae and air sacs from which the heart and the diaphragm are aerated (fig. 80). Against its dorsal wall and also covering the heart is a mass of fat cells (fig. 82 A, *FtCls*).

The Ventral Diaphragm— The ventral diaphragm (fig. 78, *vDph*) is a sheet of muscular tissue, transparent when fresh, stretched across the ventral part of the abdominal cavity above the nerve cord. Anteriorly it begins in the thorax, where it is attached on the composite endosternum of the mesothorax and metathorax. It goes back through the propodeum and petiole (fig. 53 G) and expands in the abdomen, where it is suspended by lateral points of attachment on the anterior marginal apodemes of the sterna. Beyond sternum VII, according to Morison (1928a), the ventral diaphragm

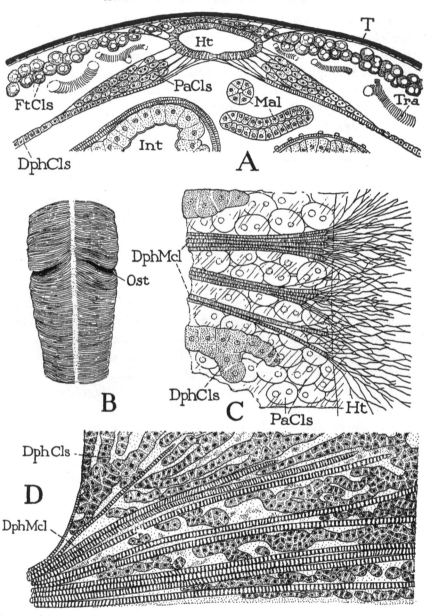

Fig. 82. Structural details of the heart and dorsal diaphragm.
A, cross section through dorsal part of abdomen, showing heart and attached diaphragm and fat cells of dorsal sinus. B, part of a heart chamber with ostia. C, part of diaphragm attached to heart. D, details of lateral part of diaphragm. For explanation of abbreviations see page 216.

212

of the worker and queen becomes much narrowed and forks into two long prongs attached on the anterior processes of the spiracular plates of segment VIII. In the drone, however, the diaphragm ends in segment VI above and in front of the last abdominal ganglion. The muscles of the ventral diaphragm radiate from their points of attachment on the sterna, but they form a compact sheet of anastomosing fibers, most of which run transversely from one side to the other. They are somewhat slenderer than the fibers of the dorsal diaphragm, but they have the structure of other body muscles with axial nuclei. The ventral diaphragm has no cellular elements and appears to have no supporting membranes. When the abdomen is contracted, the diaphragm is thrown into wide transverse folds, and in specimens freshly killed with an anesthetic it may often be seen pulsating with strong wavelike movements running posteriorly.

The Ventral Sinus— The ventral sinus, or space beneath the ventral diaphragm, communicates with the abdominal cavity above it by the lateral openings along the borders of the diaphragm between the attachment points of the latter (fig. 78). The abdominal nerve cord traverses the upper part of the sinus above a ventral layer of fat cells. The ventral commissures of the tracheal system with their saclike swellings cross the sinus from one side to the other between the nerve cord and the fat cells. All these parts may be seen very clearly through the transparent diaphragm.

Accessory Pulsating Organs— In addition to the heart and the diaphragms, there are present in many insects accessory pulsating organs that assist in maintaining the flow of the blood, particularly through the antennae and the wings. In the bee a pulsating organ in the head was noted first by Janet (1911) and more fully described later by Freudenstein (1928). It is a thin-walled vesicle lying close against the front wall of the head between the bases of the antennae (fig. 79 A, *hv*), from each side of which is given off a tubular vessel that follows the antennal nerve (*Nv*) through the antenna to the tip of the latter. According to Freudenstein, the walls of the vesicle are not muscular, but they are attached by connective tissue strands to the muscles of the pharynx. The vesicle is thus expanded by the contraction of the pharyngeal muscles, and blood is thereby drawn into it through an opening in its posterior wall (not seen in the figure). The elastic walls of the vesicle then con-

213

tract and drive the blood through the lateral vessels into the antennae. The returning blood stream, Freudenstein suggests, is probably drawn out of the antennae by the suction of blood streams in the head flowing past the antennal bases.

In the mesothorax of the bee Freudenstein describes a muscular membrane stretched beneath the air sac in the hump of the scutellum, which he regards as a pulsating organ for drawing the blood up from the bases of the wings into the space around the air sac and for discharging it through an opening in the membrane into the abdomen. The blood returning along the sides of the thorax from the head enters the wing bases from below.

THE BLOOD AND ITS CIRCULATION

The hemolymph of the bee's blood, whether of a larva, the queen, a worker, or a drone, is described by Fyg (1942) as a clear, colorless or weakly yellowish liquid. The blood corpuscles, or hemocytes, he says are colorless, nucleated cells of only one kind, though of changing form, which in the living condition show slow movements of their own. In winter bees a cubic millimeter of blood contains on an average 21,000 blood cells. Metalnikoff and Toumanoff (1930) distinguished two sizes of blood cells in the bee larva. The smaller cells are 8 to 11.36 microns in diameter and comprise about 85 per cent of the hemocytes; they have large round nuclei and the cytoplasm is strongly basophile. The larger cells, 12.7 to 15.6 microns in diameter, include 15 per cent of the cells in the blood. Both kinds of cells function as phagocytes. Injection of Chinese ink, staphylococci, or *Bacillus alvei* shows that phagocytosis begins an hour and a half after injection and reaches a maximum 24 hours later. Larvae "vaccinated" with warm emulsions of microbes were found to be resistant to what would otherwise be lethal inoculations of the same. In a paper on the body fluid of the honey-bee larva, Bishop (1923b) discusses the osmotic pressure, specific gravity, pH, oxygen and carbon-dioxide capacity, the buffer value of the blood, and the changes in these properties with larval activity and metamorphosis.

The circulation of the body liquid and the course it follows through the abdomen may be demonstrated easily by the following method, since the heart and the diaphragms continue to pulsate for some time after the other body muscles are paralyzed by an anesthetic or the body dismembered: Pin an anesthetized bee to a block of cork or paraffin and remove the top of the abdomen by making a horizontal incision all around it with a pair of small scissors. Gently pull the alimentary canal to one side so as to expose the ventral diaphragm, which will be observed pulsating strongly backward. Next cut a small hole in the top of the thorax and insert into it a drop of some stain such as carmalum in a water solution. Almost immediately the color will appear in the ventral sinus of the abdomen, where it is forced backward by the wavelike vibrations of the ventral diaphragm. From the sinus it goes upward from the lateral openings through well-defined spaces between the air sacs and the alimentary canal and particularly up wide channels against the lateral walls of each segment. It will be noted in this experiment that the blood enters the abdomen from the thorax by way of the ventral sinus only and that it has very definite courses through the abdominal cavity.

The dorsal circulation cannot be observed in this same specimen because the back is removed. Therefore, take another bee and fasten it in the same manner, but now make simply a shallow median slit through the body wall of the back so as to expose the dorsal sinus and the heart from above. Insert a drop of stain into the thorax as before. After about two minutes this will appear in perceptible amount in the dorsal sinus, very much diluted with the blood, but in sufficient amount to give white blotting paper a distinct tint. In a short time, however, the heart becomes filled with the stained blood and appears as a colored tubular band along the median line. The blood, therefore, driven up the lateral and visceral channels of the abdomen from the ventral sinus by means of the rearward contractions of the ventral diaphragm, is drawn into the dorsal sinus through the openings along its sides by the contractions of the dorsal diaphragm, which expand the sinus. Within the sinus it is driven ahead, along the sides of the heart, by the forward pulsations of the diaphragm, and soon it enters the heart itself by way of the lateral ostia to be pumped anteriorly through the aorta and finally out into the cavity of the head. From the head it makes its way backward again through the thorax and once more enters the ventral sinus of the abdomen. Thus the blood has a definite and rapid circulation through the larger spaces of the body cavity, but it also bathes all the tissues of the body and penetrates all the appendages.

215

Explanation of Abbreviations on Figures 77–82

ac, antecosta.
Ant, base of antenna.
Ao, aorta.

BC, definitive body cavity.

Cdbl, cardioblasts.
Coel, coelome.

dDph, dorsal diaphragm.
DphCls, diaphragm cells.
DphMcl, diaphragm muscles.
dTraSc, dorsal tracheal air sac.

Ecd, ectoderm.
End, endoderm.

frGng, frontal ganglion.
FtCls, fat cells.

Gng, ganglion.

Hcl, hemocoele.
Ht, heart.
hv, head vesicle.

Int, intestine.

LNv, leg nerve.

LTra, leg trachea.

Mal, Malpighian tubules.
mcl, muscles.
Msd, mesoderm.

N, notum.
Nv, antennal nerve.

OpL, optic lobe.
Ost, ostium of heart (*1Ost,* first ostium).

PaCls, paracardial cells.

S, sternum.
Scl, scutellum.
Sct, scutum.
Stn, sting.

T, tergum (*IT,* tergum of propodeum).
Tra, trachea.
TraSc, tracheal air sac.

vDph, ventral diaphragm.
Vent, ventriculus.·
VNC, ventral nerve cord.

XI

THE FAT BODY, URATE
CELLS, AND OENOCYTES

THE spaces among the organs in the body cavity of most insects are largely filled with masses of a soft tissue, which usually spreads out also in flat sheets against the body wall and may invade the appendages. The loosely aggregated masses of this tissue are known collectively as the *fat body*, because the most conspicuous feature of the cells is the inclusion in the cytoplasm of globules of an oily liquid that can be demonstrated to be of a fatty nature. The cells, however, perform other important functions besides that of storing fat; in the larval stage of many insects they retain glycogen, which is consumed during metamorphosis, and in the pupal stage they elaborate protein substances that appear in the cytoplasm as "albuminoid" granules, which are discharged into the blood to be used as food for the developing adult tissues. In the Hymenoptera small cells associated with the fat cells during the pupal stage contain urate crystals, and these cells are distinguished from the others as *urate cells* or *excretory cells*. Functionally the insect fat body is an organ for the elaboration and conservation of reserve food material, but it has no true organization, its cells being individual units in no way dependent on one another.

Dispersed among the cells of the fat body in the bee and other Hymenoptera are other usually larger cells characterized by a uniformly granular cytoplasm and a yellowish color. These cells are the *oenocytes*, literally "winecells," so named because those first

217

observed had a pale-yellow color, but in some insects they have other tints. Most of the vital organs of an insect have functional counterparts in other animals, but the fat body and the oenocytes appear to serve some special physiological purposes in the insects, since they are not present even in other arthropods.

THE FAT BODY

The cells of the fat body are derived in the embryo from the lateral parts of the mesoderm. In most insects all the fat tissue is produced from the outer, or somatic, mesoderm layers, but in the bee and other Hymenoptera, according to Nelson (1915), only that part of the fat body lying in the dorsal sinus is derived from the somatic layers, the rest being formed from the inner, or visceral, layers. Writers who have contributed most to our knowledge of the fat cells and their functions in the bee and other Hymenoptera include Kocshevnikòv (1900), Terre (1900), Anglas (1901), Berlese (1901), Pérez (1903), Koehler (1921), Bishop (1922, 1923a), Schnelle (1923), Nelson (1924), Vejdovsky (1925), Schmieder (1928), and Oertel (1930).

In a very young larva of the honey bee the fat cells, as described by Nelson (1924), are few in number as compared with later stages and differ but little from the cells of the embryonic mesoderm from which they were derived. At a slightly later stage the larval fat body consists of many small polygonal cells closely pressed together (fig. 83 A, FtCls). The cytoplasm contains small globules of a yellowish oily liquid, which becomes black when treated with osmic acid or a dark brownish red when stained with Sudan III, either test showing that the globules are drops of fatty oil. The cells, therefore, are already performing their function of producing and storing fat. When pieces of the fat body that have been stained are crushed and treated with ether, the exuding oil droplets can be seen under the microscope first to decrease in size and then suddenly to explode and disappear in solution. At the end of two or three days the fat cells have multiplied and have so increased in size that the fat body has nearly attained its final relative bulk.

In the mature larva the fat body is a loose mass of lobes and branching strands occupying most of the space within the body cavity surrounding the alimentary canal. The white color of the larva is due to the density and whiteness of the fat tissue pressed against the transparent skin. The cell masses have such definite shapes and regular contours that they appear to be enclosed in an enveloping sheath, though no limiting membrane can be distinguished. In fresh specimens cell boundaries are difficult to see, but staining brings out the cell walls (fig. 83 B) and shows that the finely granular cytoplasm is now almost filled with pale-yellowish fat globules. The nuclei are irregularly oval in cells with relatively little fat, but in others they may be distorted and pressed into various shapes by the fat globules surrounding them. The analyses by Straus (1911) on the chemical components of the honey-bee larva have shown that the worker larva accumulates a large quantity of fat in its body during growth, the amount increasing from 0.04 milligram on the second day to 6 milligrams at the end of the larval stage, a fat content which is more than 17 per cent of the dry weight (34.6 milligrams) of the body of the mature larva. In the drone larva, according to Straus, fat does not appear until the third day and then continues to increase to the end of the sixth day, or for a short period after the capping of the cell.

In the final stage of the larva when the latter is sealed in its cell of the comb and, after having eaten the last remnant of its food, has enclosed itself in a cocoon, most of the fat cells are detached from one another and are floating free in the blood. The cells now become more regular in shape and assume oval, elliptical, or spherical forms. Their cytoplasm is still filled with oily globules, but in it there are now to be seen also other small granular inclusions that are not colored by fat stains or by iodine. These small inclusions are the so-called albuminoid granules, the proteid nature of which is shown by the strong pink color they take when treated with Millon's solution. The granules appear first around the nucleus. Bishop (1922, 1923a) contended that the albuminoid bodies are produced in the cytoplasm of the fat cells by granules of basophile material discharged from the nuclei, which act supposedly as enzymes in modifying the cell cytoplasm. The nuclear granules, Bishop says, are allowed to pass into the cytoplasm by a dissolution of the nuclear membrane, which is later re-formed. Schnelle (1923), on the other hand, does not agree with Bishop as to the nuclear-induced origin of the albuminoid granules; the latter he says have an acid reaction toward stains, while the chromatin granules are basophile. Bishop notes, however, that the nuclear granules change abruptly on enter-

219

ing the cytoplasm in their reaction to nuclear dyes and that the albuminoid granules become increasingly acidophile. By whatever means the albuminoid granules may be produced, their formation is an important part of the function of the fat body cells in relation to the metamorphosis of the insect, since they will be discharged into the pupal blood as nutrient material for the growing imaginal tissues.

The fat cells of the mature larva contain in addition to their fatty and protein inclusions a large amount of glycogen, as may be shown by staining with iodine. The stain appears as blackish areas in the cytoplasm, in which the fat globules remain as clear vacuoles (fig. 83 C). Straus (1911) showed that there is a rapid accumulation of glycogen in the body of a worker larva from the second day, when the glycogen content is only 0.08 milligram, to the sixth day, when glycogen has increased to 11.5 milligrams, or over 33 per cent of the dry weight of the body. Glycogen, however, as shown by Wigglesworth (1942) in the mosquito larva, may be widely dispersed in other tissues than the fat body, being present in the central nervous system but most abundantly in the muscles.

During the propupal stage of the honey bee, when the insect is still enclosed in the larval cuticle, the fat cells are mostly floating free in the blood, which to the naked eye now appears as a thick, granular, creamy liquid that fills the body cavity. Some of the fat cells are still like those of the final larval stage, being filled with large and small fat globules (fig. 83 D, 1), though the albuminoid granules have increased in numbers. Other cells contain only a few oil globules (2, 3), but in these cells the albuminoid granules (Alb) are most numerous, and some cells (4) are almost filled with them. If the cells are crushed, the granules scatter in the surrounding liquid or adhere in small masses, but they do not float about so freely as the liberated oil droplets.

It is in the pupa that the reason for the storage activities of the larval fat cells is finally revealed. Heretofore the cells have hoarded fat, protein, and glycogen; now they lavishly dispense their contents into the blood. According to Schnelle (1923), the albuminoid granules increase in size during the pupal stage from 1 micron to 12 or 15 microns. Then, when the cells are filled to the limit, the cell membrane begins to shrink and soon ruptures, allowing the entire

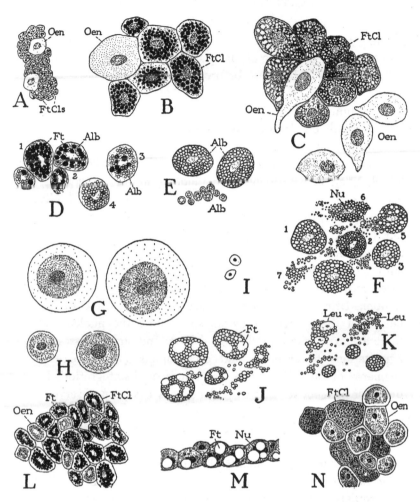

Fig. 83. Fat cells and oenocytes of worker from young larva to adult.

A, fat cells and oenocytes of very young larva. B, oenocyte and fat cells of old larva, fat cells filled with oily fat globules. C, same, showing glycogen in fat cells. D, fat cells of early propupa with albuminoid granules in cytoplasm. E, same of young pupa. F, same of later pupal period, cells in all stages of disintegration, liberating contents into blood. G, oenocytes of young pupa, free in blood. H, oenocytes of later pupal period. I, hemocytes. J, fat cells of nearly mature pupa, containing only oily fat globules. K, remains of larval fat cells persisting in young imago. L, fat cells and oenocytes of foraging worker in spring. M, section of dorsal fat body of young adult. N, fat cells and oenocytes of overwintered bee in April.

Alb, albuminoid granules; *Ft,* fat globules; *FtCls,* fat cells; *Leu,* blood cells; *Nu,* nucleus; *Oen,* oenocytes; *1, 2, 3,* accumulation of albuminoid bodies in fat cells; *4, 5, 6, 7* cells in stages of disintegration.

cell content to flow out into the body cavity, where it will become building material for the newly developing imaginal tissues.

During the early part of the pupal period, just after the larval cuticle has been shed but before the pupal eyes have begun to darken, many of the fat cells may be seen to be completely filled with albuminoid granules (fig. 83 E, *Alb*), though others still contain small oil droplets of varying sizes. At this stage all the fat cells are very fragile and are easily ruptured by the weight of a cover glass, spreading their contents through the surrounding liquid. Even in a drop of fresh blood examined without a cover glass, the plasma is seen to be full of free albuminoid bodies and droplets of oil. In such samples, though many of the fat cells are intact (F, *1*), others lack a part of the cell wall (*2, 3, 5*) or are entirely denuded (*4*). Some of the latter preserve the original cell outlines, but others (*2, 5*) have their contents streaming from ruptures. Still others (*6*) appear to be cells in a state of natural dissolution, being merely formless masses of grains and oil globules held in a thin matrix of cytoplasm around the nucleus. Finally, there are everywhere in the blood large and small masses or loose aggregations of granules and oil droplets (*7*), apparently the remnants of fat cells in the last stages of disintegration. No evidence of phagocytic action by blood cells has been observed in the destruction of the fat cells.

By the time the pupa has reached the age when the eyes have turned brown, the blood is still charged with larval fat cells and their disintegrating fragments. The majority of the fat cells, however, now have ragged outlines, and the blood plasma is filled with oil droplets and albuminoid granules, some of which are adhering in small masses but most of them floating free. Clearly the progress of disintegration in the fat cells has advanced since the earlier pupal stage when the eyes of the pupa were still pale. It is now to be noted that the oil droplets are relatively more numerous than the albuminoid granules, which have decreased in numbers. When a brownish tint begins to color the thorax of the pupa, the number of fat cells in the blood has greatly decreased, and albuminoid granules are almost absent. In the final phase of the pupal development, when color has spread to the abdomen, fat cells are still less abundant in the blood, and the albuminoid bodies have entirely disappeared, evidently having been dissolved in the blood to be consumed by the developing adult tissues. Staining with Sudan III, however, shows that fat is abundantly present both as free oil droplets of varying size and as globules in persisting cells and cytoplasmic masses. At the end of this period, just before the shedding of the pupal cuticle, the fat cells still in the pupal blood are little more than masses of fat globules. The nuclear walls have disappeared, and the cytoplasm is filled with droplets and vacuoles of oily liquid (fig. 83 J, *Ft*). So fragile are the cells at this stage that they cannot be studied under a cover glass, and by the end of the pupal period most of the larval fat cells will have ended their existence.

The young adult worker bee that has shed the pupal cuticle but which is not yet liberated from the brood cells of the comb has an ample amount of blood in the abdomen, but the plasma is a thin, clear liquid much different in appearance from the creamy fluid that filled the body cavity of the pupa. It still contains a few small fat cells (fig. 83 K) and free droplets of oil. In the fully matured adult the blood is a clear brownish or yellowish liquid containing hemocytes but no detritus or other matter except that normally dissolved in it. The fat body of the mature worker consists mostly of thin layers of cells spread against the body wall of the abdomen, especially in the dorsal and ventral blood sinuses. A less well-defined band of fat tissue lies along each side of the abdomen. The cells are much smaller than those of the mature larva. The origin of the imaginal fat cells apparently has not been definitely observed, but most writers, including Anglas (1901), Pérez (1903), Schnelle (1923), Schmieder (1928), and Oertel (1930), say that the fat cells of the adult are formed from persisting larval cells that were not destroyed in the pupa. None of these writers, however, mentions the method of multiplication of these cells; Schmieder states that there is no mitotic division of the fat cells in Hymenoptera except in the recently hatched bee larva.

The cells of the fat body in a foraging worker contain an abundance of fat, as may be demonstrated by staining with Sudan III (fig. 83 L). In young summer bees the fat content of the cells is so plentiful that the cells become swollen with large globules, which distort the nucleus (M) and may almost divide it between them. Fat cells of bees examined by the writer in January contain only a small amount of fat, but they are densely filled with granules having the appearance and staining properties of the albuminoid bodies of late larval and early pupal fat cells. In the spring the fat cells of an overwintering worker are filled with dark masses of these granules (N). Koehler (1921), in her study of the changes in the fat cells during the adult life of the bee, finds that the fat cells of wintering bees in January contain a large amount of albuminoid material. By spring, however, the albuminoids have mostly disappeared from the cells, being present in only a small percentage of the bees, and during summer they are absent entirely. At the end of September, in old bees that have ceased to act as nurses for the brood, albuminoids again appear in the fat cells. From these observations Koehler concluded that albuminoids produced in the nursing bees are elaborated into brood food as fast as they are formed in the cells but that after the nursing season they begin to accumulate in the fat cells as food reserves for the winter.

THE URATE CELLS

The urate, or excretory, cells are sparsely distributed among the cells of the fat body in the larva and the pupa of the honey bee but are not present in the adult. They are distinguished from the fat cells by their usually smaller size and by the presence of minute refractile crystals in their cytoplasm. The inclusions are presumably uric acid salts.

The urate cells of the mature honey-bee larva are described by Nelson (1924) as occurring in the abdomen interspersed singly here and there among the fat cells, with which they are always closely associated. Schnelle (1923) says the excretory cells are first seen in a 3-millimeter-long larva of the bee, either isolated in the body or attached to a lobe of the fat body. The cells change but little in structure during development, and at the end of the pupal stage they degenerate and disappear. The origin of the urate cells has not been observed, but the cells are probably derived from the mesoderm along with the fat cells and the blood cells. Some writers regard them as modified fat cells; Schmieder (1928) says the urate cells of the sawfly *Pteronidea* are derived from leucocytes that become imbedded in the fat cells during larval life.

Functionally the urate cells appear to be storage bodies for nitrogenous excretory matter. Their disappearance at the end of the pupal stage coincides with the beginning of functional activity in the newly formed imaginal Malpighian tubules. The significance of the urate cells, as stated by Schnelle, is that in the larva they supplement the work of the Malpighian tubules and in the pupa they replace the tubules. After the final ecdysis, Schmieder says, the urates in the blood are rapidly eliminated by the Malpighian tubules of the imago.

THE OENOCYTES

The oenocyte cells are derived directly from the ectoderm. In the honey bee the larval oenocytes are shown by Nelson (1915) to be formed in the embryo by proliferation from pitlike depressions of the ectoderm behind the rudiments of the spiracles on the first eight abdominal segments. From each pit a group of large cells bulges inward, some of which become detached and migrate into the body cavity as oenocytes. In a few insects the oenocytes do not separate from the epidermis, but usually they become free, though they may remain in clusters attached to the tracheae in the neighborhood of the spiracles. The oenocytes formed in the embryo of the bee persist through the larval and pupal stages, but at the end of the pupal period they are destroyed and are replaced by imaginal oenocytes newly generated from the abdominal epidermis.

In the young bee larva the oenocytes are irregularly oval, unattached cells embedded in the fat body (fig. 83 A, Oen); though relatively small, they are much larger than the fat cells (FtCls) at this stage. In the mature larva (B, C) the oenocytes have increased in size, but not so much as the fat cells, some of which are now as large as or larger than the oenocytes. Many of the oenocytes taper to a point at one end, and some have long necks thrust out between the fat cells. In the early pupa the oenocytes become large spherical cells (G) floating free in the blood. Each consists of a densely granular central body containing the oval nucleus and of a wide peripheral zone of perfectly clear substance. Later, however, the pupal oenocytes are much smaller (H) and have lost the outer clear zone. At the end of the pupal stage the larval oenocytes are said by Schnelle (1923) to be destroyed and their places taken by the much smaller, newly developed imaginal oenocytes. In the young imago a few small oenocytes are to be seen floating in the blood, but in old bees the oenocytes have compact polygonal forms and most of them are embedded in the fat cells (N, *Oen*).

Since oenocytes are almost universally present in insects and are not known in other arthropods, it is evident that they must play some important part in the physiology of insects. At present, however, the safest statement we can make about their function is that it is not known. Inasmuch as the oenocytes show an apparent secretory activity intensified at the time of the larval moults, they have been thought to have some function in connection with moulting or with the formation of the new cuticle. Also it has been suggested that they produce an oxidase enzyme, and Hollande (1914) elaborated the idea that the oenocytes are complimentary to the fat cells in that they form and conserve deposits of wax. As already noted, oenocytes seem to have something to do with the production of wax in the abdominal wax glands of the worker bee. However, since the oenocytes are present in all postembryonic stages of most insects, it is not likely that they have any single localized function.

⁂ XII ⁑

THE RESPIRATORY SYSTEM

WE ALL have to have oxygen in order to live, that is, we of the human species, the other animals, and the plants, but we do not all get it into our systems in the same way. The chief source of respiratory oxygen is the oxygen in the air and that dissolved in water. Animals and plants, therefore, can live either on land or in the water. According to which medium an animal inhabits, however, it must have a very different kind of respiratory apparatus, except insofar as one designed for aquatic life can be adapted to breathing free air, or one designed for life on land can be adapted to breathing in water. We assume that all animal life began in the water and that soft-bodied primitive animals respired by gas exchange directly through their skins. The first improvement on general skin breathing consisted of the development of specially pervious hollow folds or filaments of the skin, called *gills,* through which the blood could circulate in closer proximity to the water.

The aquatic progenitors of the land vertebrates breathed by means of gills, but for buoyancy they developed an internal air sac connected by a tube with the gullet. On leaving the water, therefore, in the manner of the modern lungfish, they could use their air sac as a lung, and to become fully land-adapted they had only to give up their gills and improve their air-sac lungs.

The early insects on leaving the water had nothing that could be converted into organs of respiration on land. Being relatively small animals, they probably got along for a while by breathing through a soft integument that had served for respiration in the water. With the acquisition of a hard, protective exoskeleton, however, the insects had to develop an entirely new way of breathing, and this they did in the simplest manner possible by the ingrowth of thin-walled branching air tubes from the body wall. Not only did the insects thus acquire a *tracheal* respiratory system, but so also did most of the other land arthropods; and, because of differences in the system in the different groups, it appears that the ancestors of each group developed their tracheae independently. Tracheal respiration, of course, is only an improved form of skin breathing. Though a tracheal system for respiration was first developed by the arthropods for life on land, it has allowed the insects at least, particularly in larval stages, to return to the water by the development of tracheated gills.

Tracheae are absent in a few small modern insects, such as most Collembola and some internal parasitic larvae, and are not functionally developed in the embryo. It has been shown by Fraenkel and Herford (1938) that various insects, especially soft-bodied larvae, can take a certain amount of oxygen directly through the skin. In blow-fly larvae ligatured at both ends to shut off the spiracles, these authors found that respiration through the integument is about one quarter of the basal amount. The earthworm furnishes a good example of a larger land animal with no respiratory organ other than its skin, but it has to live in damp places.

DEVELOPMENT OF THE TRACHEAE

The tracheal system of a modern insect begins its development in the embryo by the ancestral method of the ingrowth of air tubes along the sides of the body, and it becomes functional when the young insect leaves the egg. The embryonic tracheae contain a liquid, but usually at the time of hatching the liquid disappears and the tracheae fill with air. Sikes and Wigglesworth (1931) have suggested that the fluid in the tracheae of the embryo is absorbed into the body by osmotic pressure in the tissue fluids, which is increased at the time of hatching by the muscular activity of the young insect in its efforts to escape from the egg shell. The absorption of the tracheal liquid, therefore, draws air into the tracheae through the spiracles if the insect is freely exposed to air on hatching.

The embryonic development of the tracheae in the honey bee has been well described by Nelson (1915). The tracheal rudiments appear in an embryo about 45 hours old as a row of ten pits in the integument along each side of the body (fig. 84 A), the first pair on the mesothorax, the second on the metathorax, and the other eight pairs on the first eight abdominal segments. These pits become the *spiracles*, or breathing pores characteristic of insects. In a few insects evidence of a transient embryonic spiracle has been observed on the prothorax, and in the bee embryo Nelson found a pair of tracheal pits on the labial segment of the head. The pits on the body expand internally between the ectoderm and the mesoderm and send

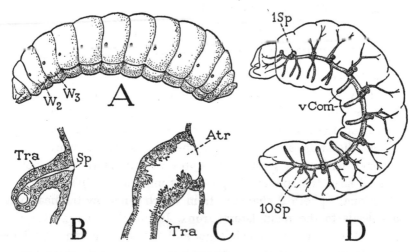

Fig. 84. Respiratory system of the larva.
A, mature larva, showing spiracles. B, section of spiracle and tracheal entrance of young larva (from Nelson, 1915). C, same of old larva (from Nelson, 1924). D, tracheal trunks and principal branches on left side of mature larva (from Nelson, 1924).
For explanation of abbreviations see page 242.

out tubular branches. Anterior and posterior branches unite to form long lateral trunks (D); ventral branches go crosswise and unite to form transverse ventral commissures (*vCom*) between the lateral trunks, and dorsal branches ramify through their respective segments. The labial pits likewise form branches which go into the head and give off posterior branches that unite with anterior branches from the mesothorax, thus extending the lateral body trunks into the head. The pits on the labial segment, however, soon close, and no trace of them remains in postembryonic stages. In the few Collembola that have a tracheal system the only spiracles are on the sides of the neck close behind the head, suggesting that these spiracles may be persisting labial spiracles.

229

Though there can be little doubt that the manner by which the tracheal system is formed in the embryo recapitulates its phylogenetic mode of origin in the early terrestrial insects, the wormlike larva of the bee or other insect must not be supposed to represent any ancestral form of the insects. When insects acquired tracheae, they had long passed the worm stage in their evolution; they were already terrestrial hexapods, as is shown by the early development of leg rudiments in the embryo, but probably they had not yet acquired wings, which develop during the larval stage.

THE LARVAL TRACHEAL SYSTEM

The tracheal system of the bee larva is relatively simple and has been fully described by Nelson (1924). It preserves the fundamental structure laid down in the embryo (fig. 84 D), and from the main trunks numerous ramifying branches go to all parts of the body. The spiracles lie in the anterior parts of their segments. They are simple openings without a closing apparatus of any kind. In a newly hatched larva the spiracular aperture is minute (B, *Sp*) and opens into a small spherical chamber, from which a narrow tracheal passage leads to the main lateral trunk. In an older larva (C) the orifice of the spiracle is much larger and opens into a more spacious atrial chamber (*Atr*) from which proceeds a wide tracheal trunk.

The bee larva, unlike the adult, makes no body movements of respiration, and consequently has no mechanical means of renewing air in its tracheae. It is probable, therefore, that oxygen diffuses into the tracheae through the spiracles as it is absorbed by the tissues from the terminal tracheoles and that a large part of the carbon dioxide produced is eliminated through the skin. Also, it is not improbable that some oxygen is taken in by cutaneous respiration.

THE STRUCTURE OF TRACHEAE

The major tracheal tubes and their branches, being ingrowths of the body wall, have the same essential structure as the integument. The epidermal layer of a trachea is a thin epithelium of flat cells (fig. 85 B, *Epth*) and is lined with a delicate cuticular intima (*In*). The tracheal intima, however, is ribbed by closely set spiral thickenings, called *taenidia* (*Tae*), that give the characteristic appearance of a trachea seen under the microscope (A). The taenidia wind around the inner wall of the trachea from right to left as followed

230

inward from the spiracles, and most of them end after making a few revolutions. The end of a broken trachea usually unravels in a narrow spiral ribbon of several parallel taenidia. In some insects, however, the taenidia of the larger tracheal trunks are circular. The taenidial thickenings of the intima give a rigidity to the tracheal walls that keeps the tubes open and maintains a free space for the passage of air.

The tracheae end in fine branches distributed over the tissues (fig. 85 C, D), and these end branches terminate with still finer

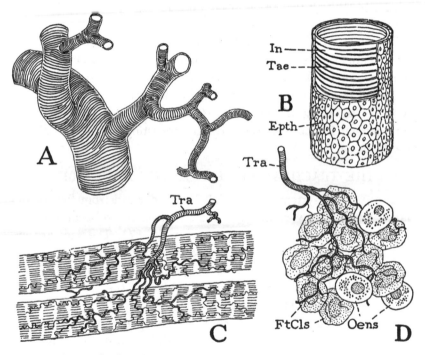

Fig. 85. Details of tracheal structure.
A, piece of branching trachea. B, structure of a tracheal tube. C, trachea and branches ending in tracheoles on muscle fibers. D, tracheae branching to fat cells, but not on oenocytes.
For explanation of abbreviations see page 242.

tubules, known as *tracheoles,* that make the final contact with the tissue cells. The tracheoles have commonly been described as lacking taenidia, but Richards and Korda (1950) report that in all species examined with an electron microscope "the tracheoles have been found to contain taenidia." It is still a disputed question whether the tracheoles end on the surfaces of the tissue cells or penetrate into the cells, but there is little doubt that it is from the tracheoles that oxygen is delivered to the cells. It has been shown by Wigglesworth (1930) from observations on mosquito larvae that the tracheoles normally contain a liquid, which, with increased muscular activity of the insect, retreats toward the ends of the tracheoles. The tracheole liquid presumably contains oxygen absorbed from the air in the tracheae. It appears, therefore, that oxygen is carried in solution through the tracheole walls wherever oxygen pressure is decreased in the surrounding tissues by metabolic activity.

The tracheal intima, being an internal continuation of the external cuticle, is largely renewed at each moult of the larva. In the longitudinal trunks the intima breaks between the spiracles, and the major tracheae and branches are drawn out through the spiracles with the shedding of the cuticle to which they are attached. The replacement intima is secreted by the persisting epithelial walls of the tracheae.

THE TRACHEAL SYSTEM OF THE ADULT BEE

The tracheal system of the adult bee is derived from that of the larva, but it is very different in general appearance because of the elaborate development of *air sacs* (figs. 86, 87). The sacs are expansions of the longitudinal tracheal trunks and of many of the branches, but students of the metamorphosis of the bee have not described their formation in the pupa. Most of the tracheae in the adult bee are more like air sacs in their structure than like ordinary tracheae, since most of them lack well-developed taenidia. They are consequently very distensible and when filled with air show as a multitude of silvery, opaque vessels, but when empty they collapse and their delicate walls are very difficult to follow in dissections. The smaller branches are so numerous in the thorax and the legs that they appear to form everywhere a glistening network among the muscle fibers and other tissues.

In the prothorax two large, tubular trunks (figs. 86, 87, *Tra*) arise from the first spiracles and converge forward through the neck into the head. Each of these trunks gives off mesally a trachea that runs posteriorly and subdivides into branches to the first legs, to the large wing muscles of the mesothorax, and to a posterior ventral thoracic air sac (5). At the base of the neck the lateral trunks are united by an anterior ventral sac (4), which splits into a pair of posterior extensions. In the head the lateral trunks end in a number of head sacs, one of which on each side (figs. 78, 86, 2) lies against the base of the compound eye and about the optic lobe, another (fig. 78, 3) is situated above the base of the mandible, and a third large dorsal sac (figs. 80, 87, 1) lies against the upper part of the face and covers the top of the brain. In the rear part of the thorax is a pair of large dorsal sacs (figs. 78, 86, 7) lying against the sides of the propodeum and connected by short tracheae with the propodeal spiracles. Above these sacs is a narrow, median transverse sac (8) occupying the cavity of the turgid mesoscutellum. The posterior ventral thoracic sac (fig. 87, 5) gives off a pair of lateral sacs (6) and tracheae to the middle and hind legs (fig. 78). The various tracheae and air sacs of the thorax finally unite in two large tubes which traverse the petiole and enter the abdomen.

In the abdomen the two tracheal tubes from the thorax expand into a pair of huge lateral air sacs (figs. 78, 86, 87, 10) that extend through the first five segments. Just behind the petiole each gives off a slender dorsal sac (fig. 78, 9). The lateral sacs are widest anteriorly, tapering posteriorly, with a constriction in each segment, and end in tracheal branches to the terminal segments. They are connected with the seven spiracles of the abdomen and are united with each other by six transverse ventral commissures (fig. 87, vCom), four of which are themselves distended into small sacs. Dorsally (fig. 86) the lateral sacs give off large segmental tracheae, which branch upon the body wall and the internal organs. Those of segments III to VI divide each into two branches (fig. 86), which enter the dorsal sinus to supply tracheae to the heart and surrounding tissues.

The air sacs have very thin walls which lack the taenidia characteristic of tubular tracheae, though the walls are roughened in most places by numerous corrugations and irregular thickenings of the intima. According to Campbell (1929), chitin cannot be detected in the air sacs and attached tracheae in either the honey bee or the house fly.

Concerning the function of the air sacs, various opinions have been expressed but none of them has been demonstrated. Where the sacs are so highly developed as in the honey bee, it would seem

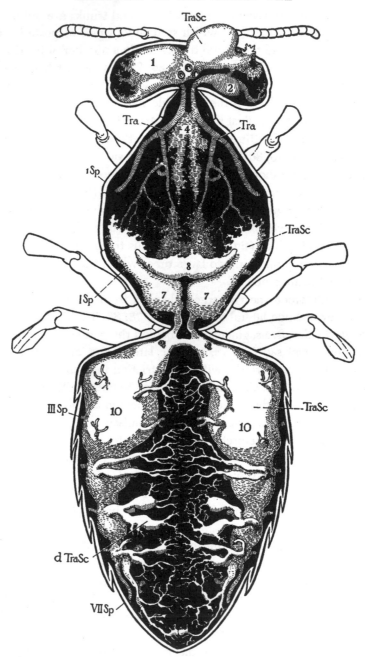

Fig. 86. Lateral and dorsal tracheae and air sacs of a worker. For explanation of abbreviations see page 242.

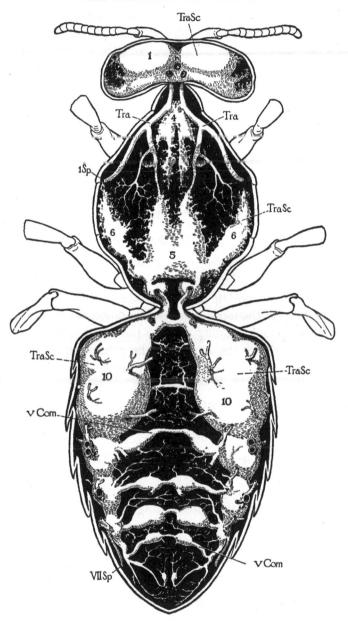

Fig. 87. General view of lateral and ventral tracheae and air sacs as seen from above after removal of dorsal tracheae and air sacs of thorax and abdomen. For explanation of abbreviations see page 242.

that they must give buoyancy to the insect in flight if they occupy space that might otherwise be filled with blood or other tissues. It is improbable, however, that the air sacs serve as storage chambers for air to be used while the insect is on the wing, when breathing might be difficult. A bee forcibly submerged in water "holds its breath" and keeps its air sacs distended but shows immediate signs of distress and becomes rigid in a very short time, showing that air in the sacs cannot serve the respiratory needs and that even an inactivated bee must have a constant renewal of air through the spiracles. If internal respiration takes place through the tracheoles, it is not likely that the air sacs serve for a direct exchange of gases with the blood, though because of their thin walls they might do so.

THE SPIRACLES

The breathing apertures of the insect are sometimes called *stigmata* (spots), but the term *spiracle* (L. *spiraculum*, a "breathing hole") is more appropriate. The apertures, however, are not always the same thing. In the more primitive condition the external opening is a primary spiracle leading directly into the trachea, but in most cases, particularly on the abdomen, the primary spiracle is sunken into a depression of the body wall, forming a spiracular *atrium*, and the external aperture is then a secondary spiracle. Associated with the primary spiracle is usually an apparatus for closing the mouth of the trachea. In the primitive condition the closing apparatus is exposed at the surface; in an atriate spiracle it lies at the inner end of the atrial chamber. Both types of spiracle structure are present in the bee, the first on the thorax, the second on the abdomen.

The first spiracle on each side of the thorax of the adult bee lies a short distance below the anterior angle of the mesothoracic wing base, but it is entirely concealed beneath the flat lobe (figs. 30, 88 A, *spl*) projecting from the rear margin of the pronotum (N_1). The edge of the lobe is provided with a dense brush of branched hairs (fig. 88 A), but when the lobe is lifted or removed (B), there is exposed beneath it a deep pocket, the front wall of which is a part of the intersegmental membrane between the prothorax and the mesothorax, and it is in this membrane that the spiracle (*1Sp*) is located. The spiracle is oval in shape, its longest diameter being about 0.14 of a millimeter in length, but it is mostly closed by a plate, or *operculum* (D, *Op*), attached along the upper half of the spiracular rim. A small arm projects from the inner end of the operculum, and on it is attached a long, slender muscle (C, D, E, *ocmcl*) arising ventrally on the mesothorax. This muscle serves to close the aperture of the spiracle by pulling the operculum down

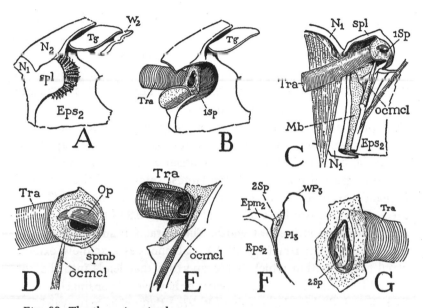

Fig. 88. The thoracic spiracles.

A, region of left first spiracle, concealed beneath spiracular lobe (*spl*) of pronotum. B, same, pronotal lobe removed exposing spiracle. C, right first spiracle and tracheal trunk, mesal, showing occlusor muscle (*ocmcl*). D, same spiracle more enlarged. E, right first spiracle, seen from within the trachea. F, left second spiracle between mesothoracic and metathoracic pleura. G, same spiracle more enlarged.

For explanation of abbreviations see page 242.

against the lower rim. It appears that the spiracle cannot be very widely opened and that the opening is never more than a crescentic slit (C) beneath the edge of the operculum. When the operculum is pulled down, however, there is nothing to hold it firm, and for this reason, apparently, despite the brush of hairs on the overlapping lobe guarding the spiracle, the mites that cause Isle of Wight disease in bees are able to get into the trachea through this spiracle. The first spiracle is a primary spiracle since the external aperture leads directly into the trachea (E), which goes forward and rapidly widens to its greatest diameter of about 0.25 of a millimeter.

The second spiracle lies in the deep membranous fold between the upper ends of the mesepimeron and the metapleuron (fig. 88 F, 2Sp). It is a minute simple opening with a thickened rim (G) and, as the first spiracle, leads directly into the connected trachea (Tra). It appears to have no closing apparatus. On account of its small size and obscure position this spiracle is difficult to find in old bees, though in young bees it is plainly seen, and figure 88 G was drawn from a worker not yet emerged from the comb cell.

The spiracles of the propodeal segment of the adult thorax are the first of the abdominal spiracles in the larva, and consequently have the structure of the other abdominal spiracles in the adult bee. The spiracles of the abdomen, as already noted, differ from those of the thorax in that the primary aperture and the closing apparatus are at the inner end of a secondary atrial chamber (fig. 89 F, Atr).

The propodeal spiracles are the largest spiracles of the bee. Each is plainly exposed on the side of the propodeum (fig. 30, IT) and presents to the exterior a long oval aperture 0.23 of a millimeter in length and 0.06 in greatest width. The spiracle is surrounded by an elevated cuticular rim (fig. 89 A) that encloses a shallow external atrial cavity, within which is the aperture that leads into the connected trachea. The tracheal aperture, however, is ordinarily closed to a narrow slit by a large valve (B, Vlv) that projects forward from within the rear lip of the elevated rim and underlaps the front lip. At B of figure 89 the left spiracle is shown with the outer rim cut off. The closing valve (Vlv) is thus clearly exposed and is seen to consist of a soft integumental fold with a strongly sclerotized margin (a). The closed valve fits into a deep groove between the outer and inner lips of the spiracle, as seen at C of the figure, which gives an inner view of the right spiracle. The ends of the sclerotic margin of the valve are produced into lobes (da and va), between which is stretched a large occlusor muscle (ocmcl). The contraction of this muscle springs the valve forward and thus closes the tracheal entrance. A second muscle (dlmcl), arising ventrally on the propodeal wall, is attached on the lower lobe of the valve and serves to open the tracheal entrance (D) by flattening the arch of the valve margin. Though the propodeal spiracles are large enough to admit two full-grown Isle of Wight disease mites abreast, no mites have been found in the tracheae leading from them, apparently for the reason that the edges of the closed valves are securely locked in the manner above described.

The other spiracles of the abdomen are located on the sides of the

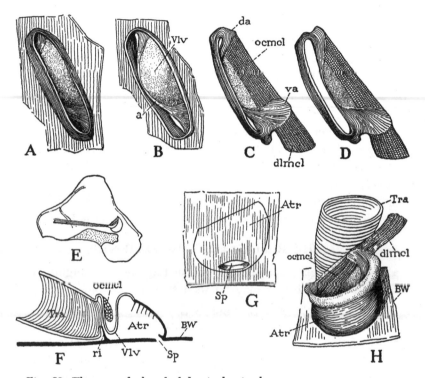

Fig. 89. The propodeal and abdominal spiracles.

A, left spiracle of propodeum, closed, outer view. B, same, atrial rim removed, exposing valve (*Vlv*). C, right spiracle of propodeum, closed, inner view. D, same, open. E, spiracle plate of segment VIII of queen, inner surface, showing spiracle and its muscles. F, vertical section through an abdominal spiracle, atrium, valve, and trachea. G, outer view of an abdominal spiracle; atrium seen through body wall. H, inner view of atrium of an abdominal spiracle, with muscles and connected trachea.

a, sclerotic edge of spiracle valve. For explanation of abbreviations see page 242.

tergal plates of the first seven segments, that is, on segments II to VIII of the primary or larval abdomen (fig. 84 A), but the last pair of the female are on the spiracular plates of segment VIII (fig. 89 E), which are concealed in the sting chamber within segment VII (fig. 59). The external openings of these spiracles are narrow slits about 0.06 of a millimeter in length, except those of the last pair, which are larger. Each opens into an atrial chamber (fig. 89 F, *Atr*), the outline of which is visible through the body wall (G). The inner wall of the atrium is covered with hairs, evidence that the atrial chamber is merely a pocket of the integument; the tracheal opening is at its inner end. With the exception of the form and size of the atrium the spiracles of segments II to VIII have the same structure as the propodeal spiracles. The tracheal entrance is similarly closed by a valvular fold (F, *Vlv*) that projects downward, with its free edge resting against a ridge (*ri*) of the body wall to which is attached the outer lip of the tracheal mouth. Within the valve fold is a closing muscle (*ocmcl*) stretched between the end processes of the valve (H). Attached on the ventral process is a long dilator muscle (*dlmcl*) from the lateral apodemal arm of the segmental sternum. Some insects have also a dorsal closing muscle attached on the dorsal process of the valve, but this muscle is not present in the bee.

THE RESPIRATORY MOVEMENTS

When a foraging bee comes in from the field and alights at the entrance of the hive, it may be seen that her abdomen is rapidly expanding and contracting, suggesting that the worker is very much "out of breath" after her recent exertions. Soon, however, the movements subside and become hardly perceptible. Various insects make respiratory movements of the abdomen; some Orthoptera breathe by a slow elevation and depression of the ventral abdominal wall, but in the honey bee the respiratory movements are principally a rapid lengthening and contraction of the abdomen accompanied by a slight dorsoventral expansion and compression.

The expiratory movements of breathing insects are always produced by muscle contraction that compresses the abdomen; the inspiratory movements, on the other hand, may result from the elasticity of the skeletal plates. The honey bee, however, as already explained, is well equipped with muscles both for extending and contracting the abdomen and for its dorsoventral expansion and compression. Lengthwise extension of the abdomen is produced by the reversed intertergal and intersternal muscles lying in the overlaps of the segmental plates (fig. 54 C, *146, 154*). As seen at D and E, a contraction of these muscles (*pmcl*) reduces the overlap between the segments and thus lengthens the abdomen posteriorly

(E). The long intersegmental muscles then act as retractors (D, *rmcl*) to pull the segments together again. In like manner the dorsoventral expansion and compression of the abdomen is caused by the reversed and direct lateral intrasegmental muscles (C, *149, 150, 151*). The first, being attached dorsally on the long lateral sternal apodemes (F, *c*), by contraction serve as dilator muscles (*dlmcl*) by depressing the sternum; the direct lateral muscles (*cmcl*) then act as compressor muscles by lifting the sternum. The two sets of antagonistic muscles are not specifically respiratory muscles since they produce the same movements of the abdomen for more general purposes.

There can be little doubt that the respiratory movements of the abdomen produce an expansion and contraction of the tracheal air sacs in the manner of vertebrate lungs. It has been demonstrated in some insects that there is a flow of air through the tracheae, at least through the lateral trunks, but experiments have not given the same results on different insects. It appears that some species inhale through the thoracic spiracles and exhale by way of the abdominal spiracles, while in others the direction of air flow is in the opposite direction.

From a study of the spiracles of the honey bee, Betts (1923) contended. that the abdominal air sacs of the bee fill with air by inspiration from the abdominal spiracles and that the compression of the abdomen then drives the air forward into the thorax and the head, whence it is discharged through the large trunks leading back to the first spiracles. Further, she suggests that decreased air pressure along the sides of the thorax in the flying bee would tend to draw the air out of the first spiracles and thus automatically create a forward circulation through the tracheal system. The first spiracles, she points out, are structurally adapted for egress of air rather than for its entrance. This theory is reasonable, but the narrowness of the connections between the abdominal and the thoracic tracheae does not seem conducive to a free passage of air currents. Brunnich (1922), experimenting on bees with cyanide, found that bees died quickly on exposure of the head and thorax alone to the fumes and that the abdomen was soon put out of action when it alone was exposed, from which he concluded that inspiration takes place through all the spiracles. Furthermore, by covering the spiracles with oil, it was found that the bees in air remained unharmed if only the

241

thoracic or the abdominal spiracles of one side were left open. Bailey (1954), from experiments on bees confined in a respiratory chamber divided into two compartments by a rubber partition around the petiole of the bee, found that air flow through the spiracles is conditioned by the activity of the bee and the amount of metabolic carbon dioxide produced. Inactive bees inhale and exhale through the first thoracic spiracles by the pumping movements of the abdomen. With active bees producing a high CO_2 tension, inhalation takes place through the first spiracles and the abdominal spiracles, and the air is exhaled by way of the propodeal spiracles. Under low CO_2 tension, however, there is but little respiration through the abdominal spiracles.

Explanation of Abbreviations on Figures 84–89

Atr, atrium of spiracle.

BW, body wall.

da, dorsal arm of spiracle valve.
dlmcl, dilator muscle of spiracle.
dTraSc, tracheal sacs of dorsal sinus.

Epm, epimeron.
Eps, episternum.
Epth, tracheal epithelium.

FtCls, fat cells.

In, tracheal intima.

Mb, intersegmental membrane.

N, notum.

ocmcl, occlusor muscle of spiracle.
Oens, oenocytes.

Pl, pleuron.

ri, ridge behind margin of spiracle valve.

Sp, spiracle (*1Sp,* first spiracle; *10Sp,* tenth spiracle).
spl, spiracular lobe of pronotum.
spmb, spiracular membrane.

Tae, taenidium.
Tg, tegula.
Tra, trachea.
TraSc, tracheal air sacs (*1–10*).

va, ventral arm of spiracle valve.
vCom, ventral tracheal commissure.
Vlv, valve of spiracle.

W, wing, or wing bud beneath cuticle.
WP₃, pleural wing process of metathorax.

❊ XIII ❊

THE NERVOUS SYSTEM
AND THE SENSE ORGANS

THE world was not made for living things; when life appeared it had to accept conditions as they were and adapt itself to them. Fortunately conditions on our planet are mostly favorable to the existence of life and to its indefinite continuance. The environment, however, is continually making changes and changing from place to place. Some of these changes will be favorable to life and others unfavorable. An animal, therefore, must have a means of orienting itself advantageously to the environment, which is to say, it must be sensitive to environmental changes and must have a responding motor mechanism. Inasmuch as the environment impinges directly on the surface of the animal, the property of sensitivity was first developed in the cells of the ectoderm. In some of the lowest metazoic animals the sensitized ectodermal cells send out nerve processes directly to the muscles, but as evolution progressed some of the sensory cells sank into the body, still retaining their muscle connections, and were then secondarily stimulated through nerves from cells that remained at the surface. Thus all the higher animals have come to have a sensory *peripheral nervous system* and a deep-seated *central nervous system*, both of which are primarily derived from the ectoderm and are still formed from the ectoderm in the embryo.

As animals have advanced in their evolution, the nervous system has assumed other duties than that of stimulating and regulating the motor mechanism according to external stimuli. Special sense organs allow the animal to react to conditions within its own tissues, motor nerves have taken on the function of regulating the secretion of glands and the discharge of hormones, and some nerve cells have acquired a secretory function. Finally, there has arisen in the central nervous system the mysterious property of consciousness, in which the messages from the sense organs are converted into sensations, which are signals of the state of the environment but convey no information as to the actual nature of the physical and chemical stimuli that give rise to them. Consciousness then becomes a new and important source of directive control in the animal by assuming a role in the activation of the muscular system independent of immediate external stimuli, but it has not yet reached a point where it can control purely physiological processes.

The essential elements of the nervous system are nerve cells, or *neurons*. Each neuron (fig. 90 A) includes the cell body, or *neurocyte* (*NCl*), and the nerve fiber, or *axon* (*Axn*), extending from it. Nutritive roots of the neurocyte are termed *dendrons* or *dentrites* (*Dnd*). A branch of the axon is a *collateral* (*Col*), and both the axon and the collateral end in fine *arborizations* (*Arb*). There are three kinds of neurons: *sensory, motor,* and *association.* In insects a sensory neuron usually has its neurocyte in or close to the epidermis (B, *SCl*), generally in connection with a sense organ, since it is the recipient of the external stimulus; its afferent axon (*SNv*) is known as a *sensory nerve.* The neurocyte of a motor neuron (*MCl*) lies in the central nervous system, and its axon (*MNv*) goes usually to a muscle (*mcl*). An association neuron (*ACl*) makes a two-way connection in the central system between a sensory nerve and the collateral of a motor nerve by an interlacing of the terminal arborizations, known as the *synapse* (*Syn*).

The environmental forces, including mechanical, thermal, chemical, and photic forms of energy, in general elicit no muscular reaction except as they effect special organs of the integument, known as *sense receptors,* or *sense organs,* which are so constructed as to be specifically receptive each to a particular kind of stimulus from the environment. The effect of the environmental stimuli on the receptor organs arouses some kind of excitation in the connected sense cells, which probably is an intensification of their ordinary metabolism, or perhaps a special kind of metabolism. This same activity, accompanied by an electrical disturbance, is then propa-

gated along the nerve to the central system, where it is transferred to the motor nerves and results in muscular movement. All
forms of environmental stimuli, therefore, are converted in the sense
cells into a *nerve impulse,* and there is no reason to believe that
there are different forms of nerve energy corresponding to the

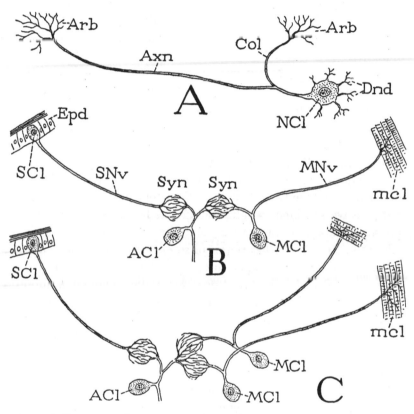

Fig. 90. Diagrams of a motor neuron and sensory-motor synapses.
A, a typical motor neuron. B, a simple sensory-motor arc with intervening
association synapses. C, a sensory nerve with double motor connections.
For explanation of abbreviations see pages 278–279.

forms of environmental energy. What the animal does depends entirely on the connections the sensory nerves make with motor nerves.
If the animal is endowed with consciousness, the nerve impulse
from an eye arouses a form of consciousness that we call light, that
from an auditory organ engenders a form of consciousness that
we call sound, not because of the nature of the stimulus but be-

cause the incoming nerve in one case goes to an optic center in the brain and in the other goes to an auditory center. In the environment there are only different kinds of physical vibration. The same is true for the other forms of consciousness, whether of smell, taste, or touch; they are simply sensations aroused by nerve impulses, the effect of which depends entirely on what part of the central nervous system the nerve enters, just as the same electric current can produce light, heat, or motion according to the nature of the apparatus to which the wire is connected.

An animal not endowed with consciousness cannot have any of the sensations that we attribute to the environmental stimuli, but the neuromuscular mechanism may work just the same, or perhaps even more precisely; error is a failure of the conscious mechanism. We do not know that insects, even the bees, have any degree of consciousness. If they have none, they are merely automatons, their actions depending on the interrelations of the sensory and motor nerves. It must be supposed then that the foraging bee does not perceive sunshine as light, though she has eyes that receive the photic vibrations from the sun, and she acts as if she sees light. Also she acts as if she has senses of smell, taste, and touch, and she reacts to changes of temperature; but all we *know* is that in response to a stimulus the insect does something. The nerve impulses from the sense organs on entering the nerve centers call forth the appropriate movements by transmitting the impulse through those outgoing nerves that will activate the proper muscles. We can understand a mechanism of this kind when we see a photoelectric "eye" open a door or do any of the other things it can be made to do, according to the apparatus with which it is connected.

The operation of a theoretically simple reflex arc, as that shown at B of figure 90, is easy to understand, but actually in the nerve centers the arborizations of numerous sensory, association, and motor neurons are intricately interwoven. If we picture a single sensory nerve making connections with only two motor nerves (fig. 90 C), it is difficult to understand how the incoming sensory impulse makes a choice of one or the other according to what action should follow the stimulus. Even we as conscious neuromuscular mechanisms do not know how we call into action the proper muscles for any desired action; in fact, we do not know what muscles we use.

When a bee flies to a flower of a particular color, she presumably does not "see" the color of the flower; the light rays coming from the flower, perhaps accompanied by odor, when transformed in the sense organs to nerve impulses, set off the motor mechanism that takes the bee to the flower. To accomplish such results the intercommunication system between sensory and motor nerves in the nerve centers must be extremely complex and quickly adjustable to different stimuli. Consider the complex action of the feeding apparatus of the bee. As soon as the tip of the tongue comes into contact with a drop of nectar or honey, the proboscis is brought into the feeding position, the tongue begins rapid back and forth movements, the sucking pump in the head becomes active in drawing the food up to the oesophagus, and probably secretion is stimulated in the salivary glands. In this act of feeding at least a dozen pairs of muscles are involved, each pair doing its own particular work, and all working in perfect co-ordination. Now let a shadow from a closely moving object fall on the feeding bee. The feeding apparatus ceases its work and is folded back under the head; the motor mechanism of the wings at once goes into action. The bee flies rapidly away; but is she "frightened" in a human sense or just automatically propelled away from a possible danger? The automatic feeding mechanism of the adult bee is perhaps a little more complex than that of the newly born infant of the human species, but the feeding act is essentially the same in both cases, and even we as adults react automatically to a sudden threat of danger.

On the other hand, the same sense organ can elicit entirely opposite motor reactions according to the quality of the stimulus. An olfactory organ, for example, stimulated by one odor may attract the insect to the source of the odor, but again another odor will have a repellent effect. In the first case the motor mechanism takes the insect to the source of the odor; in the other it takes it away. There is no evidence that different olfactory organs are sensitized to different odors, and if the nerve impulse is the same in all cases, there must be some mechanism that determines whether the reaction shall be positive or negative. Recently it has been found that there are two nerve endings in each of the sensory hairs on the tarsi and labellum of flies, and Hodges, Lettvin, and Roeder (1955) give evidence from electrical stimulation that one nerve elicits a positive response, the

247

other a negative one. How far the same may be true of other sense organs is not yet known.

Still more difficult to explain on a mechanical basis are the various complex activities of the bees in and out of the hive, things almost incredible except for the demonstrations of von Frisch (1948) and others. Consider, for example, the "dances" by which a bee communicates to other bees in the hive the direction and location of a source of nectar or pollen, and the fact that the instructed bees understand and remember the instructions and then carry them out by following a course determined by the position of the sun. Such behavior we cannot explain with diagrams of the neuromuscular system, and we can hardly resist the feeling that the bees must have some degree of intelligence. Certainly we could more easily understand their behavior if we allow them an element of consciousness, at least in its lowest form of mere sensations. Consciousness began somewhere in the animal kingdom, and why not in insects? Concerning the psychic nature of the bees, however, we know no more than they know of ours, and there is nothing to be gained by further discussing the subject in our present state of ignorance. As von Frisch (1948) says of the *geistigen Fähigkeiten* of the bees, "Über Dinge, von denen man wenig weiss, soll man nicht viel sagen."

Finally, we must note that if the working of the neuromotor mechanism of the individual bee transcends our understanding, the co-operation of many bees in their work in the hive is even further beyond our comprehension. A human builder must learn his trade, and a hundred workmen on a job must be directed and co-ordinated in their work by a supervisor. A hundred bees, however, can construct a perfect comb without either previous training or a director to co-ordinate their labor. Such actions on the part of insects we call *instinct*, but a name is not an explanation. Many insects have quite different instincts at different periods or stages of their lives, as between a larva and the adult of the same species. If instinctive actions depend on the organization of the central nervous system, it would seem, then, that the synaptic centers should be entirely reorganized as the insect takes on different habits.

In discussing the nervous system of the honey bee, we can describe its superficial structure and something of its internal organization, but of the latter much yet remains to be found out.

248

DEVELOPMENT OF THE NERVOUS SYSTEM

The central nervous system of the arthropods consists of a brain in the head and a double chain of ganglia in the body segments, the latter united with the brain and with each other by paired connectives and the two ganglia of each segment by transverse commissures. In the higher forms the two primary ganglia of each segment are combined in a single ganglionic mass, and the successive ganglia may unite in various combinations, the ganglia of the gnathal segments being commonly fused into a single suboesophageal ganglion. The motor nerves grow outward from the cells of the central ganglia; the sensory nerves grow inward from cells of the epidermis. Some writers have thought that the sense cells become nerve cells secondarily by union with outgrowing nerves, but a nerve must originate from a cell. Wigglesworth (1953) in his study of the hemipteron *Rhodnius* has demonstrated that, if a sensory nerve is interrupted, it is regenerated by inward growth of the axon, thus showing that the sense cell is a primary nerve cell. It is evident, therefore, that almost any ectodermal cell is capable of becoming a nerve cell, whether it forms part of a ganglion or remains at the surface.

The ventral nerve cords are developed in the insect embryo from two longitudinal thickenings of the ventral ectoderm, known as the *neural ridges,* with a median groove between them. The cells of the neural ridges then become differentiated into smaller outer cells that will remain as the ventral ectoderm and into larger inner cells that are the *neuroblasts,* which by further division will form the ganglion cells of the future nerve cords. A median cord of neurogenic tissue is formed over the groove between the neural ridges, and its cells contribute to the composition of the ganglia. The neural cells form at first a pair of ganglia in each segment connected by a commissure, but later the two ganglia unite in a single median ganglion. The definitive nerve cord, therefore, is a chain of segmental ganglia and interganglionic connectives. The embryonic development of the ventral nerve cord in the honey bee has been described in detail by Nelson (1915), and a full account of its formation in other insects is given by Johannsen and Butt (1941).

The brain is developed from the ectoderm of the embryonic head in much the same way as the body ganglia are generated from the ectodermal neural ridges. In the embryo of the honey bee, as described by Nelson, the areas of proliferation in the head ectoderm that give rise to the brain appear on the surface of the head as three pairs of swellings representing the protocerebral, deutocerebral, and tritocerebral lobes of the future brain. Later the swellings become less distinct, the tritocerebral swellings practically disappearing, and the contour of the head becomes relatively smooth as the brain rudiments separate from the ectoderm. The protocerebral and deutocerebral lobes of opposite sides are joined by intracerebral commissures above the stomodaeum; fibers between the tritocerebral lobes go beneath the stomodaeum and thus form a suboesophageal tritocerebral commissure. The optic lobes of the bee are not formed as a part of the primary protocerebral lobes but arise independently from involutions of the ectoderm at the sides of the protocerebrum, which then become covered by an overgrowth from the surrounding ectoderm.

THE LARVAL NERVOUS SYSTEM

The nervous system of the bee larva (fig. 91 C) includes a *brain* (*Br*) in the head above the stomodaeum, a *suboesophageal ganglion* (*SoeGng*) in the lower part of the head, and a long *ventral nerve cord* with three ganglia in the thorax and eight in the abdomen, all united by paired interganglionic connectives.

The suboesophageal ganglion (fig. 91 A, B) supplies nerves to the mandibles, the maxillae, and the labium, which are segmental appendages, and thus shows that it represents the ganglia of the three gnathal segments united in a single mass. A fourth pair of very small integumentary nerves (B, *x*) are shown by Nelson (1924) to go to the neighboring ventral epidermis. The connectives between the suboesophageal ganglion and the brain embrace the stomodaeum and are known as the *crura cerebri*, or *circumoesophageal connectives* (B, *CoeCon*).

The three thoracic ganglia of the ventral nerve cord (fig. 91 C) lie ventrally in their respective segments between the leg rudiments (*L*). Each ganglion gives off laterally two nerves on each side. The eight abdominal ganglia lie also in their proper segments; each of the first seven give off a single pair of branched lateral nerves, but the larger eighth ganglion innervates the eighth, ninth, and tenth

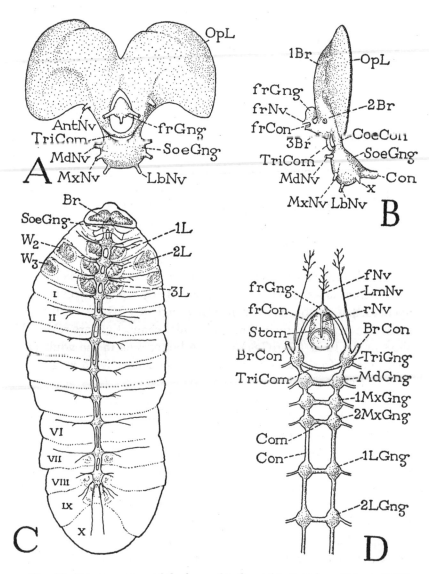

Fig. 91. Nervous system of the honey-bee larva (A, B, C from Nelson, 1924).
A, brain and suboesophageal ganglion, anterior. B, same, lateral. C, ventral
nerve cord of entire larva, dorsal. D, diagram of theoretically primitive relation
of anterior body ganglia, brain removed, dorsal.
For explanation of abbreviations see pages 278–279.

segments, showing that it includes the primitive ganglia of these segments.

The brain of the larva (fig. 91 A, B) consists of two large antero-posteriorly flattened lobes narrowly united medially with each other. The three consecutive divisions of the brain that appear in the embryo are even less distinct in the larva. The major part of the larval brain is formed of the protocerebral lobes (B, *1Br*), with the optic lobes of the future compound eyes (*OpL*) on their margins. The deutocerebrum (*2Br*) is recognizable only as a pair of slightly differentiated swellings from which arise the antennal nerves (A, *Ant Nv*). The developing antennae lie within the larval cuticle of the head. Ventrally the deutocerebral lobes merge into the tritocerebral lobes (B, *3Br*), which taper into the circumoesophageal connectives (*CoeCon*) that join the brain to the suboesophageal ganglion. The lower ends of the tritocerebral lobes, however, are bridged by a thick *suboesophageal commissure* (*TriCom*). Furthermore, the tritocerebral lobes are connected anteriorly by a pair of *frontal connectives* that go to a median *frontal ganglion* (*frGng*) on the stomodaeum in front of the brain. The frontal connectives are relatively short in the larva (B, *frCon*) and are better seen in the adult (fig. 93). Finally, from the base of each frontal connective a long nerve goes forward to the labrum.

The presence of a ventral suboesophageal commissure uniting the tritocerebral lobes of the brain strongly suggests that these lobes represent primitive postoral ventral ganglia that have secondarily been drawn upward around the oesophagus and fused with a primarily dorsal brain consisting of the ocular and antennal centers alone. The same commissure is present in most arthropods, though it may become submerged in the suboesophageal ganglion, as in the adult bee. In the insects and the myriapods there are no appendages corresponding with the tritocerebral ganglia, but in the Chelicerata and Crustacea the nerves of the chelicerae and the second antennae come from the tritocerebral ganglia; these appendages have the character of the segmental limbs and are clearly postoral in their embryonic origin. There can be little doubt, therefore, that the tritocerebral ganglia belong to a primitive first segment behind the mouth, which segment has been eliminated in the insects and myriapods following the loss of the corresponding appendages. However, embryologists have noted in some of the lower insects a trace of a segment between the antennae and the mandibles, and also in a few cases rudiments of what are taken to be transient embryonic remnants of second antennae.

The frontal ganglion is developed from the dorsal wall of the ectodermal stomodaeum shortly behind the mouth; in the adult it lies over the anterior end of the pharynx (fig. 17 B, C, *frGng*). Anteriorly the ganglion gives off a median *frontal,* or *procurrent, nerve* to the dilator muscles of the cibarium, and posteriorly a *recurrent nerve,* which runs along the dorsal wall of the stomodaeum.

If now we visualize the frontal ganglion and the tritocerebral ganglia in a supposedly primitive position respectively before and behind the mouth (fig. 91 D), then it will be seen that the tritocerebral ganglia (*TriGng*) and the gnathal ganglia (*MdGng, 1MxGng, 2MxGng*) are anterior ganglia of the ventral nerve cord and that the series of paired interganglionic connectives is continued from the tritocerebral ganglia as the frontal ganglion connectives (*frCon*). The frontal ganglion thus appears to be the first ganglion of the ventral nerve cord; but it is not developed in all arthropods, and, when absent, the connectives form a continuous loop over the stomodaeum. The primary dorsal brain, containing the ocular and antennal nerve centers, is connected with the ventral nerve cord (*BrCon*) through the tritocerebral ganglia.

THE NERVOUS SYSTEM OF THE ADULT BEE

The nervous system of the adult bee (fig. 92) is derived directly from that of the larva (fig. 91 C), and in general structure the two are essentially the same, but various modifications take place during the pupal stage. The brain particularly is more highly developed in the imago because of the presence of the compound and simple eyes and of functional antennae. The ventral nerve cord of the adult (fig. 92) has only seven ganglia as compared with eleven in the larva (fig. 91 C), resulting from the fusion of four ganglia in the thorax and of two in the abdomen. Probably extensive reconstructions take place during metamorphosis in the internal structure of the nerve centers, but no detailed studies have been made on the inner organization of the brain and the body ganglia of the honeybee larva.

The Brain— The brain of the adult bee, as seen from in front (fig. 93), differs very much in shape from the larval brain (fig.

253

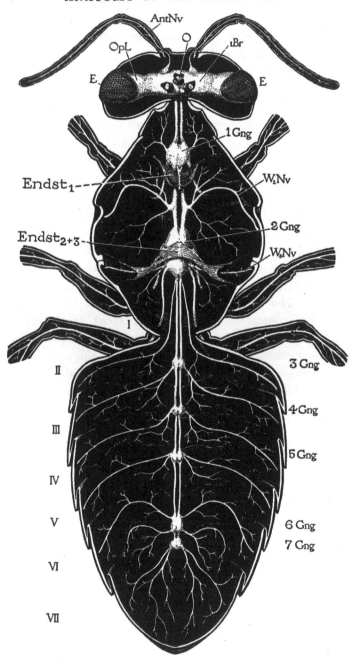

Fig. 92. General view of the nervous system of an adult worker bee, dorsal. For explanation of abbreviations see pages 278–279.

91 A), and its parts are more distinctly separated. The dorsal protocerebral lobes (fig. 93, *1Br*) are broadly joined to each other, and
the large optic lobes (*OpL*) arise from the sides of the protocerebrum by narrowed stalks. The deutocerebrum consists of two pear-
shaped antennal lobes (*2Br*) freely projecting at the sides of the
stomodaeum from the lower ends of the protocerebral lobes; from

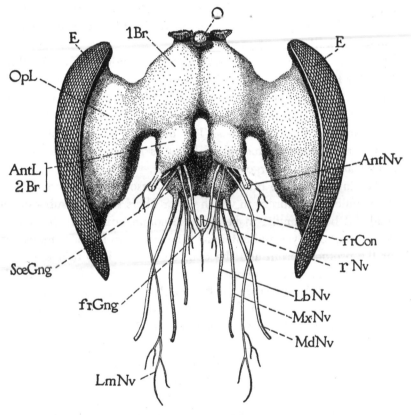

Fig. 93. Brain and suboesophageal ganglion of worker and principal head
nerves, anterior.

For explanation of abbreviations see pages 278–279.

each there is given off a large sensory antennal nerve (*AntNv*) and
smaller motor nerves to the antennal muscles. The tritocerebral
region of the adult brain lies behind the deutocerebral lobes, but
it is so reduced that it can be identified only by the origins of the
frontal connectives (*frCon*) and the labral nerves (*LmNv*), which

255

issue from beneath the deutocerebral lobes. There is no free sub-oesophageal tritocerebral commissure in the adult bee such as that of the larva, the commissure being buried in the suboesophageal ganglion (*SoeGng*). The latter, moreover, is so closely connected with the brain that the circumoesophageal connectives themselves are almost obliterated. The frontal ganglion (*frGng*) is relatively small and its brain connectives (*frCon*) are much longer than in the larva, but the ganglion gives off anteriorly a median frontal nerve and posteriorly a recurrent nerve (*rNv*) that runs back on the dorsal wall of the stomodaeum.

In its shape and size the brain differs considerably in the three castes of the bee. The brain of the drone as a whole is much larger than that of either the queen or the worker, but most of its bulk is due to the large size of the drone's head and to the greater development of the optic lobes. The brain itself, consisting of the protocerebral and deutocerebral lobes, is relatively smallest in the drone, and actually largest in the worker.

The interior of the brain is a mass of association cells and inter-communicating neurons in which motor fibers from the suboesophageal and body ganglia make synaptic connections with the sensory fibers from the eyes and antennae. The transverse fiber commissures of the protocerebrum and the deutocerebrum lie within the brain, but the tritocerebral commissure, as already noted, runs beneath the oesophagus. The internal structure of the brain of the honey bee is not essentially different from that of other insects; it has been elaborately described and illustrated by Kenyon (1896) and by Jonescu (1909).

The insect brain has no similarity in its structure to the brain of a vertebrate animal. The association centers are cellular and fibrous bodies within the brain tissue. Those of the protocerebrum include, besides the optic lobes, a pair of laterodorsal *corpora pedunculata*, a median dorsal *pons cerebralis*, and a deeper median *corpus centrale*. In the deutocerebrum are the antennal centers, which contain the terminal arborizations of the sensory antennal axons and the cell bodies of the motor nerves of the antennal muscles. Association axons from the optic lobes and the antennal centers go mostly into the corpora pedunculata, and outgoing tracts from the latter go to the motor centers of the antennae and to those of the tritocerebral and suboesophageal ganglia. The corpora pedunculata, therefore, are the principal co-ordinating centers of the brain.

The corpora pedunculata are a pair of mushroom-shaped bodies, which in the bee occupy a large part of the protocerebrum between the optic lobes (fig. 94 A, *Cpd*). Each body has two calyxlike ex-

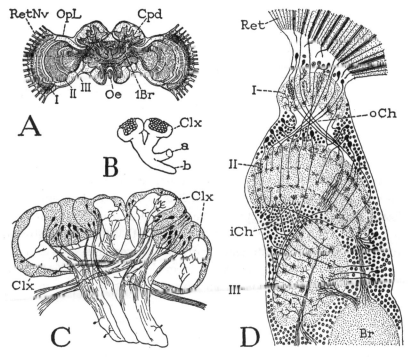

Fig. 94. Internal structure of the brain and optic lobe (A, C from Jonescu, 1900; D from Cajal and Sánchez, 1921).

A, transverse section of brain of worker. B, diagram of a corpus pedunculatum of right side, anterior. C, section of a corpus pedunculatum of a worker. D, diagrammatic section of the optic lobe and base of compound eye.

a, anterior root lobe of corpus pedunculatum; *b*, median root lobe of same. For explanation of abbreviations see pages 278–279.

pansions at the upper end (B, C, *Clx*) supported on a thick stalk that divides at its lower end into two lobes, or "roots" (B, *a*, *b*). One lobe (*a*) turns forward into the frontal region of the brain; the other (*b*) goes downward and mesally, almost meeting the corresponding lobe from the opposite side. The calyces occupy the posterodorsal region of the brain. Each calyx is a fibrous cup filled with a mass of small association cells. The axons of the cells go into the stalk, but they first give off branching collaterals into the calyces, and some fiber tracts go from one calyx to the other. Incoming nerves enter the corpus pedunculatum at the bases of the calyces (C), and others come in through the anterior root lobe. Vowles (1955) enumerates in the honey bee 10 nerve tracts in the first group and 12 in the second, both including fibers from the several synaptic centers of the optic lobes, from the sensory antennal centers, and from the suboesophageal ganglion. Nerves leave the corpus pedunculatum by way of the ventral root lobe. According to Vowles, these nerves go to the central body of the protocerebrum, to the motor centers of the antennal muscles, and to the suboesophageal region.

The corpora pedunculata are thus seen to be the principal centers through which the behavior of the insect is guided by the sensory stimuli received from the eyes and antennae. The brain is not the all-important organ to an insect that it is to a vertebrate animal; the loss of the brain merely deprives the insect of its ocular and antennal senses and thereby abolishes motor reactions that depend on stimuli from the eyes and the antennae. Many complex actions, such as those of walking and flying, may be performed automatically under the control of the body ganglia, even when the insect is deprived of its brain by having its head removed.

The Suboesophageal Ganglion— The suboesophageal ganglion (fig. 93, *SoeGng*) lies in the lower part of the head close beneath the stomodaeum and is connected with the brain by the circumoesophageal connectives. In the bee, however, these connectives are so short that the lower ganglion appears to be attached directly to the tritocerebral parts of the brain. The ganglion is a wide, flattened mass of nerve tissue, from which issue three pairs of large nerves to the gnathal appendages, namely, the mandibular nerves (*MdNv*), the maxillary nerves (*MxNv*), and the labial nerves (*LbNv*). Posteriorly the suboesophageal ganglion gives off a pair of long connectives that go to the first ganglion of the thorax (fig. 92, *1Gng*).

The Ventral Nerve Cord— The ventral nerve cord of the adult bee consists of only seven definitive ganglia and their paired connectives (fig. 92, *1Gng–7Gng*). The first ganglion (*1Gng*) lies in the prothorax above the sternal plate of this segment and in front of the prothoracic endosternum (*Endst₁*); its principal nerves go to the first pair of legs. The second ganglion (*2Gng*) is a very large nerve mass lying before the bases of the middle legs and beneath the arch of the combined endosterna of the mesothorax and metathorax (*Endst* $_{2+3}$). From this ganglion nerves go to the bases of both pairs of wings, to the middle and hind legs, and to the muscles of the mesothorax, the metathorax, the propodeum, and the first segment of the abdomen behind the petiole. The second ganglion in the thorax, therefore, is a composite ganglion formed by the union of the ganglia proper to the four segments innervated from it, which are distinct and segmentally arranged in the larva (fig. 91 C). The next two ganglia (fig. 92, *3Gng, 4Gng*) lie in segments II and III of the abdomen, but since the nerves from each ganglion go to the segment immediately following, these ganglia belong respectively to segments III and IV. The fifth ganglion (*5Gng*) lies in abdominal segment V and innervates this segment. The sixth ganglion (*6Gng*) is in segment VI, but it supplies nerves to both this segment and segment VII. This ganglion, therefore, is a composite ganglion represented by the separate ganglia of segments VI and VII of the larva (fig. 91 C). The last ganglion of the adult, lying in the anterior part of segment VII of the female bee (fig. 92, *7Gng*), represents the last ganglion of the larva (fig. 91 C), which itself is composed of the primitive ganglia of segments VIII, IX, and X. The change from the simple structure of the larval nerve cord to the condition in the adult is accomplished in the pupa.

THE SENSE ORGANS

The essential element of an insect sense organ is a sensory cell of the ectoderm with a nerve process extending into the central nervous system. Inasmuch as the impulse aroused in a sensory cell is presumably the same regardless of the nature of the stimulus, it is the structure of the cuticle part of an external sense organ that determines whether the sense cell will respond to pressure, odor, taste, sound, or light. A sense organ that receives environmental stimuli, therefore, includes a specifically selective cuticular structure, a sense cell or group of sense cells, and one or two associated cells that form the cuticular part of the organ. The whole complex is termed a *sensillum.*

Sense organs in general are classed as *exteroceptors* and *proprioceptors* according as they respond to stimuli of the environment outside the animal or to stimuli arising in the tissues of the animal itself. Insects have sense organs of each kind, but in structure their organs are so different from our own that in many cases we cannot be sure of their function. For this reason the sense organs of insects are more easily classified by their structure than according to their function. Conforming with the nomenclature proposed by Schenk (1903), the nonocular sense organs of the honey bee may be described as *sensilla trichodea, basiconica, coeloconica, ampullacea, campaniformia, placodea,* and *scolopophora.*

Sensilla Trichodea— In organs of this type (fig. 95 A) the cuticular part of the sensillum is a small hair, or seta (*Set*), set on a membrane in an ordinary setal socket over a sublying cavity in the

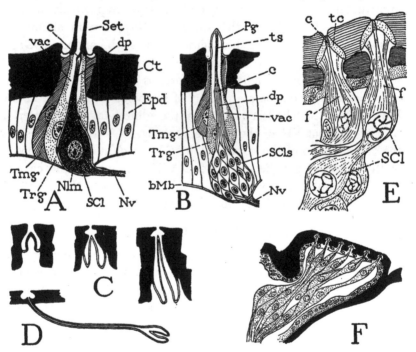

Fig. 95. Examples of sense organs.

A, diagrammatic section of a sensillum trichodeum, probably a tactile organ. B, same of a sensillum basiconicum, or surface peg organ, probably a chemoreceptor. C, examples of sunken peg organs, sensilla coeloconica. D, a sensillum ampullaceum, or Forel flask. E, section of a campaniform organ on base of hind wing of honey bee (from Newton, 1931). F, group of campaniform sensilla (from Newton, 1931).

c, cuticular connective of sense cell; *f,* axial fiber of sense cell. For explanation of abbreviations see pages 278–279.

cuticle (*Ct*). The cellular elements of the sensillum consist of three large cells of the epidermis (*Epd*). One is the hair-forming cell, or *trichogen* (*Trg*); another, the *tormogen* (*Tmg*), is the generative cell of the socket membrane; and the third is the *sense cell* (*SCl*). The sense cell has a bipolar structure: its outer end is produced into a long, slender distal process (*dp*); its inner end is continued into the sensory axon (*Nv*). The distal process may enter the seta, but usually it is attached to the base of the seta by a small conical cap (*c*), which appears to be a cuticular appendage of the setal base. In the formative stage of the organ the seta is secreted by an outgrowth of the trichogen, but later the cell cytoplasm retracts from the seta and a vacuole (*vac*) is formed in its distal part. The sense cell in some cases protrudes from the epidermis or lies entirely beneath it.

Trichoid sensilla occur on all parts of the body of the insect; in the bee they are particularly numerous on the mouth parts and the antennae (fig. 96 B, *SHr*). Most of them probably respond to movements of the seta caused by contact with some external body or surface and are therefore organs of touch. In some insects, however, it has been demonstrated that innervated hairs on the cerci respond to sound waves, and may be regarded as auditory organs, though perhaps it is going too far to say that the insect "hears" by means of them.

Sensilla Basiconica— This name is given to sense organs in which the external cuticular part has the form of a cone or peg (fig. 95 B, *Pg*). The peg may be regarded as a shortened seta, and associated with it are a trichogenous cell (*Trg*) and a tormogen (*Tmg*) as in a typical hair organ, but the sensory component consists usually of a group of sense cells (*SCls*) that send their axons into a common afferent nerve trunk (*Nv*). The distal processes of the sense cells are reduced to fine fibers (*dp*) assembled in a slender fascicle, from which a terminal strand (*ts*) extends to an attachment in the apex of the peg. The terminal strand itself is a bundle of threadlike tendons attached individually to the distal processes of the sense cells, and at each point of union there is a small dark body (*c*). The structure here suggests, therefore, that the dark body at the end of each tendon in the terminal strand represents the connective cap of the sense cell in a sensillum trichodeum (A, *c*), which in the peg organ has been drawn out at the end of a long cuticular thread.

Peg organs having multiple sense cells are generally regarded as chemoreceptors, which concept presupposes that the walls of the peg are thin enough to be permeable to the chemical stimuli of taste and smell. In the bee, peg organs are abundant on the mouth parts and on the antennae (fig. 96 B, *SPg*). According to von Frisch (1921), those of the antenna occur in transverse rows on the distal ends of the last eight subsegments of the flagellum and are most numerous on the apex. From the experiments of von Frisch (1921) there can be no doubt that the antennae are the principal seat of the bee's olfactory sense, but as yet no method has been devised for determining the function of any particular kind of sense organ situated on them.

Sensilla Coeloconica and Ampullacea— Sense organs given these names are simply sensilla basiconica in which the peg is sunken into a pit or a flask-shaped cavity of the cuticle (fig. 95 C). They occur on the antennae of the bee but are much less numerous than the organs with exposed pegs, and in some cases the sensillum resembles a hair organ in having only a single sense cell. A type of sunken organ in which the peg is contained in a swelling at the end of a long cuticular canal (D) is found in the antenna of ants and is known as a *Forel flask*. The pit-peg organs are commonly regarded as olfactory organs, but as in the case of the typical sensilla basiconica there is no way of determining their function.

Sensilla Campaniformia— The campaniform organs derive their name from the form of the cuticular part, which latter usually is a conical or domelike body (fig. 95 E, *tc*) embedded in the surrounding cuticle, into which the distal process of the sense cell is inserted like the clapper of a bell. Campaniform sensilla occur in small groups mostly on the appendages (F). Each organ has a single sense cell, which is generally represented as a simple bipolar cell with a long, slender distal process inserted into the apical cone, and is so described by McIndoo (1914b) for the honey bee. According to Newton (1931), however, the sense cell in an organ of the bee is binucleate (E, *SCl*) and is traversed by an axial fiber (*f*) that goes to the outer cone (*tc*). Neither Newton nor McIndoo has recognized any cell associated with the sense cell in these organs, but Sihler (1924) in a campaniform organ of the cockroach shows the presence of a large accessory cell, corresponding with the trichogen of a hair organ (A, *Trg*), that extends to the dome and surrounds the distal process of the sense cell. In the honey bee, Newton says, "the main central element of each sensilla is the sense-cell which itself appears to secrete non-cellular parts of the sense-organ and also to give rise to the sensory fiber." In his figure (E) the "sense fiber" (*f*) is very suggestive of the distal process of the sense cell in other organs, and it is to be suspected that the surrounding part represents the trichogen, which here forms the terminal cone (*tc*).

The campaniform organs of the honey bee are widely distributed on the mouth parts, the antennal bases, the bases of the wings, the legs, and the sting. McIndoo (1914b) has mapped them and estimated their numbers. He reports that in the drone there are 1,998 individual organs on the wing bases and 606 on the legs; in the worker 1,510 are on the wings, 658 on the legs, and 100 on the sting; the queen has 1,310 on the wings, 450 on the legs, and 100 on the sting. McIndoo called the organs "olfactory pores" because he thought the distal processes of the sense cells were exposed at the surface in small pits and were thus directly receptive to odor stimuli. Newton, however, showed that a thin surface layer of the cuticle covers the organs (fig. 95 E) and that the "sense fiber" is attached by a small refractive body (*c*), which he termed a "scolopala," embedded in the apex of the cone.

It has been pointed out by Pringle (1938a, 1938b) that the campaniform sensilla occur at places where strains and stresses are most likely to be produced in the cuticle as a result of movement. He found by means of electrodes from a recording apparatus inserted into the associated nerve that bending or pressure in the neighborhood of a group of the organs caused an electric discharge from the nerve. Pringle thus demonstrated that the campaniform sensilla are proprioceptors that respond to strains in the cuticular skeleton, and he therefore appropriately names these organs *stress receptors*.

Sensilla Placodea— In these organs the external part of the sensillum is a plate, usually flush with the surface of the cuticle (fig. 96 C, *Pl*). Plate organs are known only on the antennae and are present in Homoptera, Coleoptera, and Hymenoptera. In the honey bee it has been estimated that there are about 3,000 plate organs on each antenna of the queen, 3,600 to 6,000 in the worker, and 30,000 in the drone. The plates are elliptical in shape, from 12 to 14 microns in greatest diameter, which is lengthwise on the antenna, and are closely situated among the antennal hairs (A, B, *SPl*). Each plate is defined by a narrow marginal line of nonsclerotized cuticle (C, *a*), within which is a concentric light line formed by a submarginal groove (*b*) on the inner surface.

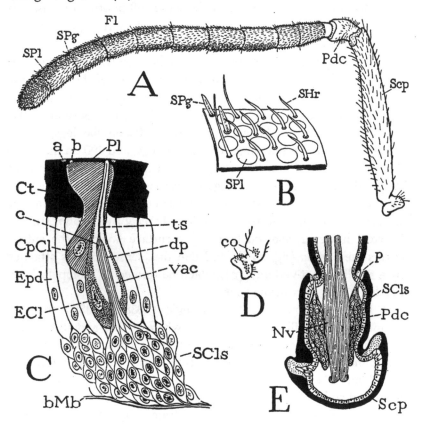

Fig. 96. Antennal sense organs.

A, left antenna of worker, showing plate organs, peg organs, and pits of organ of Johnston between bases of flagellum and pedicel. B, part of antennal surface with sensory hairs, pegs, and plate organs. C, diagrammatic vertical section of a plate organ. D, campaniform organs on base of scape of antenna. E, lengthwise section of antennal pedicel containing organ of Johnston.

a, outer ring of plate organ; *b,* inner groove of same; *c,* ends of fibers in terminal strand; *p,* pit. For explanation of abbreviations see pages 278–279.

The cellular elements of a plate organ (fig. 96 C) are the same as in a sensillum basiconicum (fig. 95 B). The numerous sense cells (fig. 96 C, *SCls*) form a compact body, from which the fiberlike distal processes (*dp*) extend outward in a slender fascicle, and are connected with the fibers of a narrow terminal strand (*ts*) arising from a point in the inner groove (*b*) of the plate. The fascicle and the terminal strand are contained in a vacuole (*vac*) of a long tapering cell (*ECl*), which is termed the *enclosing cell* but is clearly the trichogen of a peg organ (fig. 95 B, *Trg*). The neck of the enclosing cell again is embraced by a large cell (*CpCl*) known as the *cap cell*, the expanded distal end of which fills the cavity of the cuticle beneath the plate. This cell evidently is the tormogen, or socket membrane cell, of a hair or peg organ. The plate, therefore, cannot be regarded as representing the base of a hair.

The function of the plate organs is a subject on which there has been much difference of opinion, well reviewed by Wacker (1925). Some of the earlier investigators believed that these organs register vibrations; others contended that they are olfactory organs. The olfactory idea is generally accepted at present, not on a basis of experimental evidence, but because the plate organs are the most numerous sense organs on the antennae of the bee, and the antennae have been experimentally demonstrated to be the principal site of odor perception. The antennal sense organs on the flagellum occur only on the eight distal subsegments, and it has been shown by von Frisch (1921) and by Frings (1944) that, when this part is removed from both antennae, the bees give no reaction to odors on which they have been experimentally trained. If only one subsegment with sense organs is left intact on either antenna, however, the bee retains a sense of smell. Insects from which both antennae have been entirely amputated may still respond by a "common chemical sense" on exposure to the fumes of strongly irritating volatile substances to which they are not normally accustomed, and this fact has led to the erroneous conclusion that the antennal organs are not the odor receptors of the insect.

Richards (1952) has made a detailed study of the minute structure of the antennal cuticle of the honey bee, including that of the plate organs. He finds that a double-layered epicuticle extends continuously across the sense plates and the surrounding cuticle. The sublying procuticle of the plates is less sclerotized than that of the general antennal cuticle and has a complex structure in which the micelles are arranged radially in the outer parts and tangentially around the periphery. In thickness the plates measure 1 to 2 microns at the margins, 0.5 to 1.25 microns at the center. Though the plates may be supposed to be permeable to odors, the cell beneath the plate is evidently the placogen and has no direct relation to the deep-seated sensory apparatus. From experiments with electrical connections on the antennal nerve of various insects, including the honey bee, Chapman and Craig (1953) report "there was no sign, in any preparation, of a direct response to chemical stimulation." The electrical method of identifying the function of sense organs thus appears to be eliminated from the study of odor receptors, at least until a more refined technique is devised, but in any case the problem of determining which organs on the antennae are responsive to odor would still remain. The only objection to regarding the plate organs as odor receptors is their structure.

Sensilla Scolopophora— A scolopophorous sense organ, or scoloparium, consists of a group of simple sensilla, each having its individual attachment on the body wall, but the attachment cells are closely associated, not dispersed as in a group of campaniform organs. The point of attachment on the cuticle may be marked by a pit or thickening, but generally a scolopophorous organ has no differentiated external part, except where it is associated, as in some Orthoptera, with an auditory tympanum.

A typical scolopophorous sensillum (fig. 97 A) is composed of three consecutive cells. The outermost cell is the cap cell ($CpCl$) as in a plate organ, corresponding with the tormogen of a hair organ; the intermediate cell, or enclosing cell (ECl), is equivalent to the trichogen; the basal cell is the sense cell (SCl) with a long, slender distal process (dp) going through the enclosing cell to be attached by a terminal fiber (t) traversing the cap cell from its origin on the cuticle. The distinctive feature of a scolopophorous sensillum, from which it gets its name, is the presence of a hollow, sharp-pointed, rodlike structure enclosing the end of the distal process of the sense cell. This rod (Sco) is the *scolops* (*Stift* of German writers), called also the "scolopala." The sharp apex (c) of the scolops suggests the attachment cap of the sense cell in a hair organ (fig. 95 A, c) connected with the cuticle by the tendon (fig. 97 A, t) traversing the cap cell. According to Sihler (1924), Hsü (1938), and Richard (1952), the scolops is cast off with the cuticula at the moults in young insects. At the base of the scolops the distal process of the sense cell is vacuolated (vac), and an axial fiber (axf)

traverses the sense cell from the nucleus to a dark body in the end of the scolops.

In some cases the scolopophorous organs are connected by a basal ligament with the body wall, and the organs first studied were those associated with tympana in Orthoptera. Hence they were formerly called "chordotonal" organs, but in general there is nothing to suggest that they are organs of hearing. Most of them appear to be proprioceptors and perhaps register the effects of muscle tension on the body wall.

Scolopophorous organs occur in various parts of the body and in the appendages of most insects. In the honey bee they are known to be present in the head and in the legs. Janet (1911) described a pair of small, fusiform "chordotonal" organs in the head of the bee innervated from branches of the antennal nerves, each of which terminates in a cord attached to the articular membrane of the antenna. According to Debaisieux (1938), there are four scolopophorous organs in each leg of the honey bee, one being in the femur, two in the tibia (one proximal, the other distal), and one in the tarsus. The femoral organ lies close to the trochantero-femoral articulation but is attached by a long fascicle of tendons in the distal end of the segment.

The proximal organ in the tibia lies a short distance below the "knee" and is known as the *subgenual organ* (fig. 97, B, *sgO*). It has been described by Schön (1911), McIndoo (1922), and Debaisieux (1938). The organ is a conical mass of cells (C) attached by one angle of its broad distal end to the posterior wall of the tibia. McIndoo shows the connection as if formed by only a few enlarged cells of the epidermis, but according to Schön a large mass of cells spreads from the epidermis against the cap cells, and small "accessory" cells cover the ends of the latter. Probably the difference depends on the plane in which the section is cut. In the ants a large bundle of long club-shaped epidermal cells spreads against the entire end surface of the cap cells. The sense cells (*SCls*) are massed in the narrow proximal end of the organ; their long distal processes traverse the enclosing cells, and their ends are encased in elongate scolopes (*Sco*) that project into the bases of the cap cells (*CpCl*). The axons of the sense cells enter a sensory nerve (B, C, *SNv*) that joins the main leg nerve (B, *Nv*) and receives also the nerves from a group of campaniform organs (*co*) on the head of the tibia.

The distal organ in the tibia of the honey bee (fig. 97 B, *dO*) has the more usual form of a scolopophorus organ. It was described by McIndoo (1922) simply as a group of "ganglion cells," but Debaisieux (1938) showed that it is a true scoloparium connected

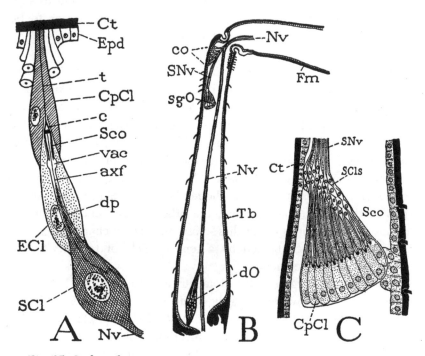

Fig. 97. Scolopophorous organs.

A, diagram of a single element in a scolopophorous organ. B, diagrammatic section of hind tibia of a drone, showing campaniform sensilla on base and subgenual and distal scolopophorous organs (from McIndoo, 1922). C, diagrammatic section of subgenual scolopophorous organ in hind tibia of a drone (from McIndoo, 1922).

t, attachment fiber of sense cell. For explanation of abbreviations see pages 278–279.

with the leg nerve. The sense cells are somewhat irregularly arranged, and their distal processes converge to a point of attachment in the distal end of the tibia. The organ in the tarsus is described by Debaisieux as a tarsopretarsal scoloparium consisting of three sensilla, two of which are attached at the bases of the pretarsal claws and one on the sheath of the unguitractor.

The Organ of Johnston— This organ, which lies in the pedicel of the antenna, was first described in the mosquito by Johnston (1855) and still bears his name. It consists of numerous elongate sensilla forming a cylinder about the axial nerves of the antenna. The sense cells send their nerves into the scape, where they join the antennal nerve trunks; their distal processes are attached in the articular membrane between the pedicel and the base of the flagellum, or to sclerotic processes of the latter. In complexity of structure the organ varies much in different insects (as well as in the descriptions of different writers) and reaches its highest development in the males of Chironomidae and Culicidae. In most insects the organ is composed of typical scolopophorous sensilla, but in the Odonata, according to Eggers (1923), it consists entirely of long-necked sense cells without scolopoid rods or associated cells.

The Johnston organ of *Vespa* as described by Child (1894) appears to be composed entirely of sense cells, the distal processes of which he says become rods (*Stäbchen*) that go to the articular membrane between the pedicel and the flagellum. Child's figure of the organ in *Vespa* closely resembles that given here for *Apis* (fig. 96 E), but Berlese (1909) figured the sensilla of *Vespa* as including both cap cells and enclosing cells as well as sense cells. McIndoo (1922) made what appears to be a detailed study of the organ in the honey bee, and he found only a mass of small sense cells with long, slender distal processes attached in the articular membrane at the base of the flagellum. It is clear that the structure of the organ of Johnston in *Vespa* and *Apis* needs further investigation.

The base of the first subsegment of the flagellum in the honey bee presents a circle of numerous minute, closely set, spinelike points radiating into the articular membrane, and the sense-cell processes appear to be directly attached to pits in the membrane at the ends of the points. This structure suggests, therefore, that the organ of Johnston is an apparatus for registering the movements of the flagellum, though in the bee the flagellum has little freedom of movement on the pedicel.

The Ocelli— The ocelli are the three small eyes situated in a triangle anteriorly on the top of the head in the worker and queen (fig. 13 A, B) but lower on the face of the drone (C). The external part of each ocellus is a thick, biconvex lens (fig. 98 G, *Ln*), which is simply a thickening of the head cuticle (*Ct*). Beneath the lens and enclosing its sides is a cellular body of several kinds of cells, all derived from the epidermis (*Epd*). Immediately surrounding the inner face of the lens is a layer of corneagenous cells that secreted the lens. Those under the middle of the lens form a thin transparent

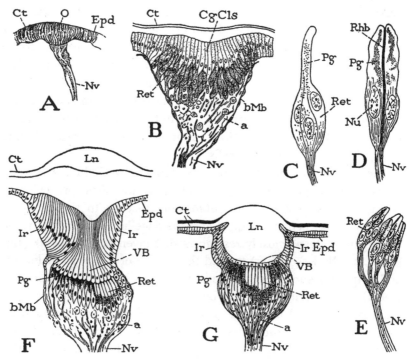

Fig. 98. Development and structure of an ocellus of the honey bee (from Redikorzew, 1900).

A, section of median ocellar rudiment in epidermis of a young pupa. B, section of lateral ocellus at later stage of pupa. C, group of retinula cells from young pupa. D, retinula cells at later stage with rhabdom formed between them. E, four retinula cell groups and nerve. F, section of lateral ocellus of old pupa (lens separated). G, section of median ocellus of adult bee.

a, interstitial cells of ocellus. For explanation of abbreviations see pages 278–279.

layer known as the *vitreous body* (*VB*); the thicker cells at the sides of the lens may be termed the *iris cells* (*Ir*). Beneath the vitreous cells is a bulb-shaped cellular mass that tapers inwardly into the *ocellar nerve* (*Nv*). An outer layer of long, parallel cells, perpendicular to the lens but separated from the latter by the vitreous layer, constitutes the *retina* (*Ret*), which is the light-sensitive

270

part of the ocellus. The retinal cells are arranged in groups of two or three and sometimes more on the ends of nerve branches (C, D, E), and each group is termed a *retinula*. The opposing surfaces of the cells in each retinula secrete an axial rodlike structure, known as a *rhabdom* (D, *Rhb*), which probably serves to deflect the light coming through the lens into the surrounding cells. The inner ends of the retinular cells are produced into nerve fibers that traverse a mass of interstitial cells (G, *a*) behind the retina and go to the brain through the ocellar nerve (*Nv*).

In its development an ocellus of the bee has been shown by Redikorzew (1900) to originate in a very young pupa from a simple thickening of the epidermis (fig. 98 A, *O*), but even at this early stage some of the cells are differentiated into optic sense cells and have axons that form an ocellar nerve (*Nv*). As the optic rudiment increases in thickness (B), the future retinal cells (*Ret*) sink into its deeper part, and those destined to secrete the lens remain at the surface as a thick layer of corneagenous cells (*CgCls*). During the secretion of the lens the corneagenous cells greatly increase in depth, and in an old pupa (F) they clearly show a differentiation into a peripheral group (*Ir*) that will become the iris cells of the adult (G, *Ir*), and a central group (*VB*) that, on completion of the lens, will be reduced to the vitreous layer of the adult. (In the section at F the lens has been detached from the corneagenous cells.)

An ocellus probably responds only to degrees of light. The lens probably does not throw an image on the retina, since its focal point would be much beyond the outer ends of the retinal cells. A dark pigment formed in the retina (fig. 98 F, G, *Pg*) has been shown to collect in the basal parts of the cells in the dark and to move into their outer ends under the influence of light.

The Compound Eyes— The structure of the compound eye of the honey bee is not essentially different from that of other insects. The eye is composed of a large number of ocular units termed *ommatidia* (fig. 99 A, *Om*), covered externally by a cuticular cornea (*Cor*). The ommatidia lie close together in the eye, but they are separated and optically isolated by intervening pigment cells; they taper inwardly and converge to the narrowed outer end of the optic lobe (*OpL*). The surface of the cornea is differentiated into six-sided facets corresponding with the outer ends of the ommatidia. The narrow rims of the facets are opaque, but the central areas are

271

transparent and constitute the lenses of the ommatidia. The number of ommatidia in a single eye, reckoned by the number of facets in the cornea, has not been exactly counted in the bee, and estimates by different writers vary considerably, but there are probably 4,000 to 5,000 in the worker, 3,000 to 4,000 in the queen, and 7,000 to 8,000 or more in the drone. The surface of the bee's compound eye,

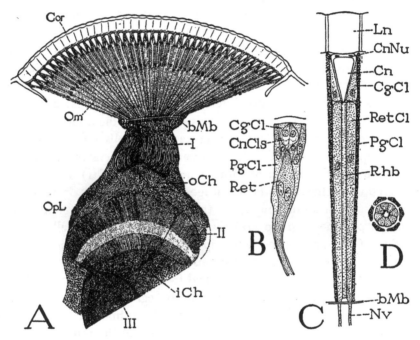

Fig. 99. The compound eye (A, B from Phillips, 1905).
A, vertical section of compound eye and optic lobe, diagrammatic. B, ommatidium of young pupa before formation of rhabdom. C, diagrammatic lengthwise section of an ommatidium. D, cross section of same.
For explanation of abbreviations see pages 278–279.

especially in young bees, is covered with long unbranched hairs arising from the rims of the facets; in old bees many of the hairs become brushed off. The purpose of these eye hairs is not known; Phillips (1905) has shown that the base of each hair is penetrated by a long binucleate hair-forming cell, but the cells have no nerve connections, showing that the hairs are not sensory organs.

In a vertical section of an ommatidium (fig. 99 C) the lens (*Ln*) is seen to be an elongate prism only slightly convex on its outer and inner surfaces. Succeeding the lens is a *crystalline cone* (*Cn*) about as long as the lens, with its apex directed inward. The cone is formed of four cells, but most of the cell cytoplasm becomes converted into a transparent hyaline substance in order to transmit the light from the lens, leaving remnants of the nuclei (*CnNu*) at the distal end. The cone cells are enclosed in two pigmented corneagenous cells (*CgCl*). In insects that have compound eyes in their immature stages these cells at each moult come together over the cone and secrete the new lens; in apterygote insects and crustaceans which moult throughout adult life, the corneagenous cells are retained permanently between the lens and the cone.

Proximal to the cone is the light-sensitive part of the ommatidium, composed of a fascicle of eight or sometimes nine long, slender cells (fig. 99 C, *RetCl*), which together constitute a retinula. The retinula cells are the optic sense cells; their inner ends are produced into the retinal nerves (*Nv*), which penetrate the basement membrane of the eye (*bMb*) and extend into the optic lobe. In the axis of the retinula is a long, rodlike rhabdom (*Rhb*) formed by the opposed inner surfaces of the retinula cells. The entire ommatidium is enclosed in pigment cells (*PgCl*), which in some insects comprise a distal group in the region of the cone and a proximal group around the retinula.

The retinula cells have a striated appearance on their inner borders, where they come together to form the rhabdom. From the work of Hesse (1901) on arthropod eyes, the striae of the retinula cells were formerly regarded as the enlarged ends of nerve fibrils traversing the cells from the nerve, and it was supposed that they fuse to form the rhabdom (see Weber, 1933, Snodgrass, 1935). The rhabdom accordingly was thought to be the light-sensitive part of the eye. Hesse, however, expressed the opinion from the results of staining that the striations of sensory cells sometimes take on a cuticular structure. Later, convincing reasons for believing that the rhabdom is a cuticular secretion of the retinula cells were given by Nowikoff (1931a), and Machatschke (1936) demonstrated that the rhabdom reacts positively to staining tests for chitin. Since the retinula cells are primarily epidermal cells, they should retain an ability to secrete a cuticular structure. The probable function of the rhabdom, therefore, is to scatter the vertical rays of light entering the eye into the sensory cells of the retinula. Nowikoff

(1931b) devised a simple apparatus for condensing light from a conical reflector into the end of a narrow glass tube. He found that, if the tube is filled with water, the light traverses it and throws a spot of light from its distal end. On the other hand, when the tube is filled with fine glass rods suspended crosswise on a thread (in imitation of the striae of the rhabdom of the eye), the light is radiated from the sides of the tube.

In its development the compound eye arises from a thickening of the ectoderm over the surface of the optic lobe. In a newly hatched larva of the honey bee, according to Phillips (1905), there is yet no differentiation of the cells that will form the future eye. During later larval life some of the cells become associated in spindle-shaped groups (fig. 99 B, *Ret*), which are to be the retinulae. The inner ends of these cells send out nerve processes into the optic lobe, which, as shown by Nelson (1915), has already been formed in the embryo. The retinula cells later withdraw from the cuticle, and the other cells close over them to generate the lens. These primary corneagenous cells then become themselves differentiated into groups: four above the axis of each retinula (B, *CnCls*) form the cone by conversion of their cytoplasm into a vitreous substance, and two cells surrounding the cone (*CgCl*) remain as the corneagenous pigment cells of the adult. Long peripheral cells (*PgCl*) become the outer pigment cells that enclose the cone and the retinula. At the beginning of the pupal stage of the bee, Phillips says, the ommatidia are completely formed.

A comparison of the compound eye with an ocellus shows that the principal anatomical difference between the two pertains to the cornea. In the ocellus there is a single corneal lens for the entire eye; in the compound eye the cornea is broken up into numerous small lenses, one for each group of retina cells. A minor difference between the two types of eye is the presence of a vitreous cone in each ommatidium of the compound eye and of a single vitreous body or layer of vitreous cells in the ocellus. It is evident, therefore, that the compound eye is not a group of ocellar eyes, since an ocellus is much more complex in its structure than is an ommatidium.

It is well known that bees recognize objects with their compound eyes, that they perceive movement and distinguish colors and forms, but we do not know how visual stimuli are registered in their nervous system. It is particularly difficult to understand how colors are distinguished, since there is no known mechanism in the insect eye corresponding to the rods and cones of a vertebrate eye. The rhabdom disperses whatever wave lengths of physical light it receives into the retinular cells, and with unpolarized light all the cells must be stimulated alike. Different wave lengths might be supposed to give different degrees of stimulation, but not different kinds of stimulation, and it seems doubtful if degrees of stimulation corresponding with wave lengths of light would be perceived as different colors. Yet experiments leave no doubt that bees distinguish colors by sight, at least from the red into the ultraviolet. Physiologists, however, cannot explain how colors are registered in the human brain. The perception of form by the compound eyes is perhaps less difficult to understand, since the ommatidia diverge in all directions from the optic lobes and a pattern of the visual field might be registered in them. The compound eye, however, seems best constructed for the perception of movement, and most insects are quickly aware of moving objects.

The insect compound eye with its cylindrical light receptors has one potential that our eye does not have; this is its ability to register the plane of polarized light. That bees "see" a pattern of polarized light and can guide their flight by it has been amply demonstrated by von Frisch, and the manner by which the polarization of light is reproduced in the eye appears to have been well explained by von Frisch (1950), Autrum and Stumpf (1950), and Menzer and Stockhammer (1951). Polarized light so called is light vibrating in one plane. When it is directed lengthwise through a glass cylinder filled with a light-scattering liquid, the cylinder viewed at right angles to the plane of vibration appears illuminated, seen in the opposite direction it is dark (see Bragg, 1933, fig. 92). The rhabdom of the compound eye, therefore, should have the same effect on polarized light as the glass cylinder. In ordinary light the rhabdom scatters the rays to all sides and thus stimulates all the cells of the retinula. Polarized light, on the other hand, will be deflected most strongly into the two retinular cells that lie opposite the vibration plane, and but weakly or not at all into the others. The bee in this manner receives in her compound eyes a "picture" of the polarized light.

That the incidence of polarized light in the eye must be registered as a particular pattern has been demonstrated by von Frisch (1950) with an "artificial eye" made of eight triangular pieces of polaroid fitted together like the sections of the rhabdom. Sunlight seen through this "eye" produces the identical pattern that would be expected: two opposing sections are bright, those at right angles are dark, and the intermediate sections gray. In the eye, of course, the effect of the light is not illumination but stimulation of the retinular cells. Each cell, having its own individual nerve, is a functional unit; consequently an impulse is transmitted to the brain only from those cells of the retinula that most strongly receive the light from the rhabdom.

That the insect compound eye should be a natural analyzer of polarized light is evident from its structure, but how the bees utilize this property of the eye for guiding them to a source of food when the sun is obscured by clouds is more difficult to understand. The only necessary condition is that there must be somewhere in the heavens a patch of clear sky; the light from the sky is partly polarized. It is well known that when the sun is visible a bee returning from a feeding place performs a "dance" in the hive, the direction of which, relative to the perpendicular on the comb, gives to the bees following her the angle that the direction to the food source makes with the direction of the sun. The outgoing bees remember their instructions, correctly judge the proper angle of flight, and arrive at the feeding place.

If the day is cloudy but still some blue sky is somewhere visible, the bees are able to direct their flight by the polarization of light from the sky. With his "artificial eye" von Frisch demonstrated that polarized light makes a distinctive pattern for every part of the sky depending on the time of day. Since the plane of vibration is perpendicular to the direction of the sun, the position of the stimulated segments in the "eye" gives an index to the direction of the sun. The bee, by taking note of the particular pattern of stimulation in her retinulae, can thus orient herself at any time of the day to the position of the sun. This assumes a time sense on the part of the worker bee, which she is known to have, but her mental method of calculating the position of the sun from the polarized pattern in her eyes is her own secret. Yet, that bees can and do use polarized light from the sky as a compass is well attested by the work of von Frisch, and Thorpe (1949), after working in the field with von Frisch, reports that the evidence is conclusive. Furthermore, the returning forager orients her dance in the hive so that her followers are able to guide themselves by the polarization pattern from the sky to the food source. However, von Frisch admits that an unsolved problem still remains; even when the sky is totally obscured by clouds, the bees somehow find their way with a close degree of accuracy to an indicated feeding place. Thorpe concludes that in spite of the long association between man and bees we are still far from understanding the behavior of the bees and their social organization. It may be added that as anatomists we are even further from an understanding of the sensory and nervous mechanisms by which the bees are able to do things in a way that we ourselves can scarcely understand.

The Optic Lobe— The optic lobe of the compound eye, formed as a simple involution of the ectoderm in the embryo, attains an extremely complex inner structure in the adult (fig. 94 D). Structurally and functionally the optic lobe of the insect eye corresponds in some respects to the retina of a vertebrate eye. The lobe is a mass of nerve fibers and association nerve cells and characteristically presents three distinct internal synaptic regions, the outermost (*I*) termed the *lamina ganglionaris*, the next (*II*) the *medulla externa*, and the third (*III*) the *medulla interna*. Between the first and the second, and the second and the third, some of the nerve fibers from opposite sides cross each other, forming an *outer chiasma* (*oCh*) and an *inner chiasma* (*iCh*). Most of the retinal nerve fibers end in the lamina ganglionaris, where they form synapses with association cells, the axons of which go into the medulla externa or the medulla interna. Some of the retinal nerves themselves, however, extend into the medulla externa, where they form synapses with nerves going into the medulla interna, and finally from the latter nerves proceed into the brain. The diagram (fig. 94 C) represents only a fraction of the intercommunicating neurons of the optic lobe; how such a complex apparatus transmits specific nerve impulses engendered by light entering the retinal cells and transmutes them into motor reactions, or into a perception of light intensities, color, form, and motion, is still beyond our ability to comprehend.

Explanation of Abbreviations on Figures 90–99

ACl, association nerve cell.

AntL (2Br), antennal lobe of brain, deutocerebrum.

AntNv, antennal nerve.

Arb, arborization of nerve ending.

axf, axial fiber of a sense cell.

Axn, axon.

bMb, basement membrane.

Br, brain (1Br, protocerebrum; 2Br, deutocerebrum; 3Br, tritocerebrum).

BrCon, brain connective.

CgCls, corneagenous cells.

Clx, calyx of corpus pedunculatum.

Cn, crystalline cone of compound eye.

CnCls, cone cells.

CnNu, nucleus of cone cell.

co, campaniform organs.

CoeCon, circumoesophageal connective.

Col, collateral branch of an axon.

Com, nerve commissure.

Con, nerve connective.

Cor, cornea.

CpCl, cap cell of sense organ.

Cpd, corpus pedunculatum.

Ct, cuticle.

Dnd, neural dendrites.

dO, distal scolopophorous organ of tibia.

dp, distal process of sense cell.

E, compound eye.

ECl, enclosing cell of sense organ.

Endst, endosternum; Endst $_{2+3}$, composite endosternum of mesothorax and metathorax.

Epd, epidermis.

Fl, antennal flagellum.

Fm, femur.

frCon, frontal ganglion connective.

frGng, frontal ganglion.

frNv, frontal nerve.

Gng, ganglion.

I (fig. 94), lamina ganglionaris of optic lobe; II, medulla externa; III, medulla interna.

iCh, inner chiasma of optic lobe.

Ir, iris cells.

L, leg.

LbNv, labial nerve.

1LGng, 2LGng, ganglia of first and second leg segments.

LmNv, labral nerve.

Ln, lens.

mcl, muscle.

MCl, motor nerve cell.

MdGng, mandibular ganglion.

MdNv, mandibular nerve.

MNv, motor nerve.

1MxGng, first maxillary ganglion; 2MxGng, second maxillary ganglion.

MxNv, maxillary nerve.

NCl, nerve cell, neurocyte.

Nlm, neurilemma.

Nu, nucleus.

Nv, nerve.

O, ocellus, or ocellar rudiment in epidermis.

oCh, outer chiasma of optic lobe.

Oe, oesophagus.

Om, ommatidium of compound eye.

OpL, optic lobe of brain.

Pdc, antennal pedicel.

Pg, sense peg, pigment.

PgCl, pigment cell.

Pl, plate of plate sense organ.

Ret, retina.
RetCl, retinal cell.
RetNv, retinal nerve.
Rhb, rhabdom.
rNv, recurrent nerve.

SCl, sensory nerve cell; *SCls,* cells.
Sco, scolops.
Scp, scape of antenna.
Set, seta.
sgO, subgenual scolopophorous organ.
SHr, sensory hair.
SNv, sensory nerve.
SoeGng, suboesophageal ganglion.
SPg, sensory peg.
SPl, sensory plate.

Stom, stomodaeum.
Syn, synapse.

Tb, tibia.
tc, terminal cone.
Tmg, tormogen.
Trg, trichogen.
TriCom, tritocerebral commissure.
TriGng, tritocerebral ganglion.
ts, terminal strand of attachment fibers of sense cells.

Vac, vacuole of sense organ.
VB, vitreous body of an ocellus.

W_2, W_3, wing rudiments.
WNv, wing nerve.

✣ XIV ✣

THE ENDOCRINE ORGANS

WE HAVE seen in the last chapter that the nervous system regulates most of the activities of the animal, such as muscular movement and gland secretion. Other vital processes, however, go on in the animal that the nervous system does not directly effect, such things, for example, as metabolism, digestion, growth, and development. These are the more fundamental properties of living matter; they begin in the egg and the embryo before the neuromuscular system is developed and are controlled by chemical substances called *enzymes* and *hormones*. Hormones are produced in special *endocrine glands*, also in secretory cells of the nervous system, and are dispersed directly into the blood. The hormones of insects that are best known, though principally by their effects, are those that regulate growth, moulting, and developmental changes of structure. It must be understood that hormones have nothing to do with determining the *course* of development; the form and structure that the insect takes on during its development are determined by the hereditary *genes* present in the chromosomes of the egg nucleus.

Most animals grow continuously from youth to maturity, but an insect with a hard nonexpansible integument cannot grow in this manner; it must cast off its inelastic cuticle at intervals and each time form a new and larger one that briefly permits a resumption of growth and development before the new cuticle hardens. An animal that grows in this intermittent fashion, therefore, must have some provision for arresting its growth processes between moults. In the young insect this inhibition of growth is effected by a hormone secreted in a pair of glandular bodies in the head, known as the *corpora allata* (fig. 100 A, *Ca*), which lie against the oesophagus

280

shortly behind the brain. If this so-called *juvenile hormone* were present in effective amount in the blood at all times, however, the insect could never become an adult. Hence at each growth period the juvenile hormone loses its potency or is counteracted by another hormone that causes the insect to moult and rapidly but briefly promotes its growth and development.

The growth and development hormone comes primarily from neurosecretory cells in the pars intercerebralis of the brain. It is passed out through nerves to a pair of cellular bodies behind the brain termed the *corpora cardiaca* (fig. 100 A, *Cc*), from which it is discharged into the blood, perhaps with secretions from the cardiacum cells themselves. The results of many experiments have seemed to show that the brain secretion acts finally on a pair of *thoracic glands* lying in the ventral part of the prothorax and that the effective hormone of growth and development ultimately comes from these glands of the thorax. The neurosecretory centers of the brain, the corpora cardiaca, and the thoracic glands, therefore, have been regarded as integral parts of a complex endocrine system responsible for the moulting, growth, and development of the young insect. It has been found by Chadwick (1955), however, that nymphal cockroaches from which the thoracic glands have been removed will nevertheless moult and develop into normal reproductive adults. Further discussion of the subject is best left to the specialists on hormones.

Inasmuch as the wings of insects become functional only in the adult stage, the mature insect at once acquires a greatly increased sphere of activity in its environment. It can leave its restricted juvenile habitat and take up ways of living and of feeding quite impossible to the wingless young. Consequently the young of many insects have likewise gone their own way in the world, choosing new habitats and new sources of food, and, in becoming structurally adapted to their new ways of living, have entirely lost the parental form. The young honey bee, for example, has no resemblance to its parents; it is a specialized larval form well adapted to being reared in a cell of the hive comb. No matter how specialized the larva may be, however, it must still carry the potentiality of undergoing a final transformation into the reproductive adult. The corpus-allatum hormone maintains the larval form, whatever it may be, up to the end of the larval span of life; then this juvenile hormone weakens

281

and the growth and development hormone takes complete control. The insect either changes at once into the imaginal form, or it does so in an intermediate pupal stage. Shortly after transformation the thoracic glands disappear, but the secretory centers in the brain remain active, and both the corpora allata and cardiaca persist into the adult.

NEUROSECRETORY CELLS OF THE BRAIN

Hormones are now known to be produced by neurosecretory cells in various parts of the nervous system in most groups of animals, both vertebrates and invertebrates. A comprehensive review of the subject may be found in the paper by Scharrer and Scharrer (1954) on the hormones of neurosecretory cells. Among insects the principal seat of neurosecretion is in the pars intercerebralis of the protocerebrum, but in addition, according to the authors just mentioned, neurosecretory cells are present in the suboesophageal ganglion, the ventral ganglia, and the frontal ganglion of many species.

Neurosecretory cells in the brain of the honey bee were perhaps first observed by Weyer (1935), who described them as glandular nerve cells containing granules of secretion, best developed in the worker, less so in the queen, and least in the drone. Though it appears that little further attention has been given to secretory cells in the brain of the honey bee, Thomsen (1954) has given a full account of neurosecretion in other adult Hymenoptera, including Sphecoidea, Vespoidea, and Apoidea. In this order, according to Thomsen, there is a single median group of neurosecretory cells in the anterior dorsal part of the pars intercerebralis of the brain, though in most other insects there is also a lateral group on each side. The cytoplasm of the secretory cells contains numerous dark-staining granules of secretion matter, but the nuclei resemble those of nerve cells, and among the secreting cells are larger nonsecretory nerve cells. The axons of the secretory cells proceed downward and posteriorly through the brain as a single fiber tract, which divides into two strands that emerge as a pair of nerves which go to the corpora cardiaca. In most insects the nerves from the median cell group cross each other before leaving the brain (fig. 100 A), and Thomsen observes that at least some of the fibers in the Hymenoptera form a chiasma. In insects having lateral groups of secretory cells, the axons from these cells form a second pair of nerves to the corpora cardiaca (*ccNv*). The secretory activity of the brain cells is shown by the fact that granules of secretion formed in them can be traced into the axons and followed along the latter into the corpora cardiaca.

THE CORPORA CARDIACA

In general, the corpora cardiaca are oval bodies lying on the stomodaeum shortly behind the brain (fig. 100 A, *Cc*) at the sides of the hypocerebral ganglion (*hcGng*), with which they are connected by short nervos. The aorta, however, runs between the corpora cardiaca, and the latter are usually closely appressed against the aortic walls, although there is much variation in their actual position and in some insects the corpora cardiaca are united with each other over the aorta.

The association of the corpora cardiaca with the aorta has given these bodies their name, but they are developed from the dorsal wall of the stomodaeum in connection with the hypocerebral ganglion of the stomatogastric nervous system. In earlier papers on the nervous system, the bodies now known as the corpora cardiaca were called the "pharyngeal ganglia." Thus, under this name, Roonwal (1937) described the origin of the corpora cardiaca in the embryo of the locust from cells differentiated from the posterior end of the median rudiment of the hypocerebral (occipital) ganglion. Pflugfelder (1937) in like manner described the bodies in the phasmatid *Dixippus* as formed by cells of the stomodaeal wall at the sides of the hypocerebral ganglion, though they later become closely attached to the aorta. The development of the corpora cardiaca from the stomodaeal wall leaves no doubt that these bodies are ectodermal, and their close association in origin with the hypocerebral ganglion suggests that their cells were originally nerve tissue.

The corpora cardiaca of the honey-bee larva have been described by Schaller (1950) and by L'Hélias (1950). They are so little developed in the larva, however, that they have generally been overlooked, and L'Hélias says it is impossible to see them in dissections. In microtome sections they appear as small loose masses of cells (fig. 100 D, *Cc*) closely attached to the walls of the aorta. In the bee larva the aorta (*Ao*) is greatly enlarged behind the brain and is open below, so that in sections it appears as an arch lying over the

283

oesophagus (*Oe*) and the recurrent nerve (*rNv*) on the dorsal wall of the oesophagus. As described by L'Hélias, the corpus cardiacum of the bee larva consists of a few loosely connected large cells and interspersed smaller cells. The large cells resemble secretory nerve cells such as those of the brain. The single nerve of each body arises from two roots in the brain (*Br*); the fibers traverse the corpus cardiacum and continue into the corpus allatum (*Ca*) of the same side, but some of them unite with the recurrent nerve. Nelson gives a figure (1915, fig. 42) of a transverse section through the head of a newly hatched larva of the honey bee which is very similar to that by L'Hélias (fig. 100 D). He shows lateral masses of cells in the walls of the aorta, but he did not recognize them as being the corpora cardiaca of the larva.

Considering the importance of the corpora cardiaca in the endocrine control of pupal metamorphosis and adult development in most insects, it is surprising that the organs should be so poorly developed in the bee larva. In sawfly larvae of the family Diprionidae the corpora cardiaca are shown by L'Hélias (1952a) to have the usual postcerebral position against the walls of the aorta. Each body is connected with the brain by a single nerve trunk with two roots in the pars intercerebralis.

The corpora cardiaca of adult Hymenoptera are described by Thomsen (1954) as oviform or pear-shaped bodies composed of large and small cells. In the honey bee they are a pair of small bodies (fig. 100 B, *Cc*) lying on the dorsal side of the oesophagus (*Oe*) just where the latter joins the pharynx (*Phy*) in the occipital foramen of the head. The aorta (*Ao*) runs between them. According to Hanan (1955), the corpora cardiaca of the adult honey bee are united above the aorta and are connected with each other by a dorsal commissure going over the recurrent nerve and by a ventral commissure beneath the oesophagus. The cells of the cardiac bodies in *Apis* are said by Thomsen to be small, highly vacuolated, and without distinct borders. Among the Hymenoptera in general only a single nerve goes to each corpus cardiacum from the brain, but it is observed to arise from two roots, suggesting that it contains the fibers of the usual two nerves in other insects. The fibers are said by Thomsen to diverge into the corpus cardiacum and to go between the cells; they contain dark-staining granules derived from the secretory cells of the brain and "show many large swellings which

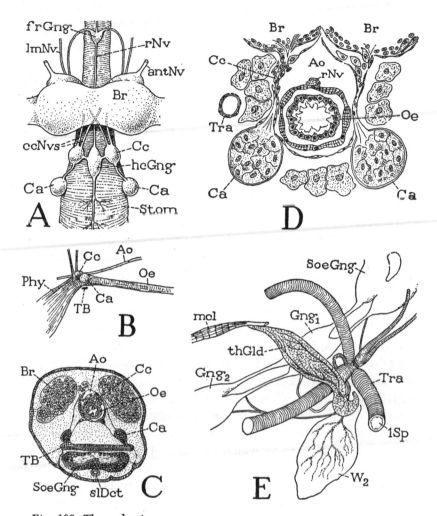

Fig. 100. The endocrine organs.

A, diagram of general position and relations of the retrocerebral endocrine organs of insects. B, retrocerebral endocrine organs of adult honey bee. C, section of head of bee embryo, corpora cardiaca cells generated from dorsal wall of oesophagus, corpora allata arising at ends of crossbar of tentorium (from Pflugfelder, 1948). D, corpora cardiaca and allata of honey-bee larva, seen in cross section through anterior end of aorta and oesophagus (from L'Hélias, 1950). E, right thoracic gland and associated structures of honey-bee larva (from L'Hélias, 1952).

antNv, antennal nerve; *Ao,* aorta; *Br,* brain; *Ca,* corpus allatum; *Cc,* corpus cardiacum; *ccNvs,* nerves of corpora cardiaca; *frGng,* frontal ganglion; *Gng₁,* *Gng₂,* first and second thoracic ganglia; *hcGng,* hypocerebral ganglion; *lmNv,* labral nerve; *mcl,* muscle; *Oe,* oesophagus; *Phy,* pharynx; *rNv,* recurrent nerve; *slDct,* salivary duct; *SoeGng,* suboesophageal ganglion; *1Sp,* first spiracle; *Stom,* stomodaeum; *TB,* tentorial bridge; *thGld,* thoracic gland; *Tra,* trachea; *W₂,* rudiment of hind wing.

285

are fusiform, ovoid or spheroid in shape." The secretion from the brain appears to be stored for some time in the corpora cardiaca, finally to be discharged through the surfaces of the organs into the blood stream. Most of the neurosecretory material, Thomsen says, "collects along the medial surface of the corpus cardiacum, from where it may ooze out into the aorta." Further evidence suggests that the corpora cardiaca are also themselves secretory organs, but little is known of their function in adult insects.

THE CORPORA ALLATA

The corpora allata of most insects are globular or oval cellular bodies lying against the stomodaeum (fig. 100 A, *Ca*) at the sides of or behind the corpora cardiaca, with which they are connected by short nerves. They are developed, however, in the ventral part of the head of the embryo from the ectoderm of the first maxillary segment and secondarily migrate to their definitive positions. The corpora allata are thus "transposed bodies," as their name implies. Pflugfelder (1937) very clearly illustrates their origin in the phasmatid *Dixippus* as cellular ingrowths between the bases of the maxillary and mandibular rudiments. In the honey bee, according to Nelson (1915), the corpora allata are formed as tubular ingrowths of the ectoderm of the first maxillary segment close behind the bases of the adductor apodemes of the mandibles. The cavities soon disappear in the rudiments, which become solid cellular masses and are then constricted off from the body wall and carried dorsally. In the honey-bee larva the corpora allata are relatively large globular bodies (fig. 100 D, *Ca*) easily seen in dissections lying at the sides of the oesophagus and connected with the corpora cardiaca by nerves continued from the latter. Each is composed of large polyhedral, strongly eosinophile cells enveloped in a thin membranous tunic.

The corpora allata of adult Hymenoptera are described by Thomsen (1954) as globular or oviform organs connected by nerves to the lower or posterior parts of the corpora cardiaca. The corpusallatum nerve is a direct continuation of some of the fibers in the corpus cardiacum, which extend through the latter from the brain to go into the corpus allatum, where the nerve divides into branches that ramify among the cells. Stained granules of secretion matter can be followed along the nerve fibers, and a few may be seen among the cells, though most of the neurosecretory material seems to disappear in the corpora allata. In the adult honey bee the corpora allata (fig. 100 B, *Ca*) lie against the sides of the oesophagus close to the corpora cardiaca, with which they are connected by very short nerves.

Little experimental work has been done on the hormonal control of metamorphosis in the honey bee. Schaller (1952) reports that ligaturing a last-stage larva behind the head results in the production of a form within the larval cuticle that is much closer to a mature pupa than to a propupa. The precocious development is most pronounced in the integument, but it affects also the alimentary canal and the muscles. The pupal stage, Schaller says, has been jumped over; the ligatured worker larva passes directly from the larval stage to that of an imago almost ready to emerge. These experimental results probably mean that the juvenile hormone from the corpora allata was shut off and that enough of the developmental hormone from the corpora cardiaca had already diffused into the blood to bring about adult development. In other experiments Schaller (1951) found that the time of development in the ligatured worker larva is shortened by four days, which gives it the same developmental period as that normal to the queen, and yet the typical worker characters are formed.

Though the corpora allata are of particular importance in the young insect because they furnish the hormone that maintains the juvenile status up to the adult or the pupa, they persist into the imago and appear to take on other functions. Pflugfelder (1948) has shown that during the larval period of the honey bee the corpora allata approximately double their size in each of the five larval stages, each body in the final stage being about 30 times its size in the embryo. In the pupa there is at first a sudden decrease in size, followed by a small increase, which is lost again in the young adult worker. In midsummer or overwintering workers, however, the corpora allata increase again to seven times or more their size in the young imago. In the queen they increase only five times during adult life, and in the drone only three times. It has been thought that the chief function of the corpus-allatum hormone in the adult insect is to stimulate secretion in the accessory sex glands of the male and the production of yolk in the ovaries of the female. Pflugfelder points out, however, that if the secretion of the corpora allata in the adult honey bee is specifically a gonadotrophic hormone, the bodies should show the greatest increase in size in the queen, whereas actually the increase is far greater in the worker. He argues that the corpus-allatum hormone of the adult must be a regulator of the general metabolism of the insect, which would be of greatest intensity in the worker because of the greater amount of work she performs.

THE THORACIC GLANDS

Thoracic glands, presumably of an endocrine nature, have now been found in preimaginal stages in representatives of most of the larger orders of insects and therefore probably serve some important function in the juvenile life of the insect. In the larvae of higher Diptera corresponding glands are not present in the thorax, but it is thought that they enter into the composition of the complex endocrine ring gland. Although these glands are commonly termed "thoracic" glands, they are formed in the embryo from the ectoderm of the second maxillary segment, and in mayflies and phasmatids they appear to remain in the head as a pair of small ventral head glands.

The thoracic glands of the honey-bee larva have been described as follows by L'Hélias (1952b). They lie at the sides of the alimentary canal on the line between the prothorax and the mesothorax close to the first spiracles. Each gland (fig. 100 E, *thGld*) is elongate fusiform in shape, transparent and slightly opalescent, and is composed of many small, closely packed cells. From the neighboring tracheal trunk the gland receives a branching trachea which ramifies among the cells, and it is innervated from the suboesophageal, prothoracic, and mesothoracic ganglia. With the beginning of pupation the glands commence to degenerate, and slowly they disappear.

Inasmuch as hormones play a very important part in the regulation of general physiological activities in the higher animals, it seems probable that insects should have hormones yet to be discovered in addition to those that govern juvenile growth and metamorphosis. Undoubtedly the life processes of insects, as those of other animals, are activated by innumerable enzymes. The living animal is fundamentally a chemical organization; its skeleton, muscles, and nerves are mere adjuncts that have been added for mechanical purposes in the course of evolution. The more complex an animal becomes, however, the more finely adjusted to its increasingly mechanical organization must be its chemical effectors and regulators. The evolution of an animal, therefore, has involved the development of intricately related chemical constituents along with the development of physical structure, and if natural selection has been the only guiding influence its task has been far greater and vastly more complex than simply that of originating species.

THE REPRODUCTIVE
SYSTEM

THE organs that compose the reproductive system have themselves nothing to do with reproduction. They simply house and nourish the germ cells in their development into spermatozoa and ova and furnish passageways that serve for the extrusion of the gametes to the exterior. The male insect usually has external organs of copulation and intromission; the female is provided with a chamber outside the actual body cavity in which the eggs are held for insemination by spermatozoa stored in a connected sperm receptacle, and in many insects she has an ovipositor for placing the eggs in some particular place. The external genitalia, however, are highly variable in different insects, especially in the male, and an egg-laying organ may be entirely absent or, as in the bee, converted into a sting.

The primary germ cells of insects, as explained in the first chapter, are probably in all cases derived from cleavage cells in the early development of the egg, though in the honey bee their early origin has not been traced and in the young embryo they are not distinguishable from the mesoderm cells of the forming ovary or testis. In the bee the somatic cells of the blastoderm of a fertilized egg develop into a female adult, and the germ cells become eggs; in an unfertilized egg, however, the somatic cells form a male bee and the germ cells develop into spermatozoa. The result is, therefore, that after maturation the male cells remain haploid in their chromosome numbers, while after fertilization the female cells are diploid.

THE MALE ORGANS

The reproductive organs of the male honey bee (fig. 101) include a pair of *testes (Tes)*, their outlet ducts, or *vasa deferentia (Vd)*, which are partly enlarged as *vesiculae seminales (Vsm)*, a pair of huge *mucus glands (MGld)* united at their posterior ends, the single *ductus ejaculatorius (Dej)*, and a large intromittent organ, or *penis (Pen)*, the last being a complex internal structure, which is everted at the time of mating. The testes, the vasa deferentia, including the seminal vesicles, and the mucus glands are formed in the embryo from the mesoderm and at this stage have no connection with the exterior. The ejaculatory duct, on the other hand, is a tubular ingrowth of the ectoderm that meets the rear ends of the mucus glands and eventually opens into them. The primitively external opening of the ejaculatory duct is the *male gonopore*. The penis, finally, is formed as a second ingrowth of the ectoderm that carries the gonopore inward with it, so that the ejaculatory duct comes to open into the anterior end of the penis lumen. The external aperture of the inverted penis may be distinguished as the *phallotreme* (fig. 103, *Phtr*).

The Testes— In a mature drone the testes (fig. 101, *Tes*) are small, flat, triangular bodies of a yellowish color overlying the anterior ends of the mucus glands. Each testis has an outer tunic within which are numerous tubules opening into a common chamber at the end of the vas deferens. The larval testes (fig. 102 A) are described by Nelson (1924) as a pair of elongate reniform bodies about 3.75 millimeters in length lying at the sides of the heart in the fourth, fifth, and sixth abdominal segments. They are composed of numerous parallel strands of cells representing the definitive tubules, bound together by connective tissue. From the rear end of each testis proceeds a very delicate rudiment of the vas deferens *(Vd)*. The development of the testes has been followed in detail by Zander (1916). The organs attain their greatest development in the pupa; between the fifth and the sixth day of the pupal period spermatogenesis is completed, and by the seventh day the testes occupy a large space in the posterior part of the abdomen. According to Bishop (1920a), four days before the emergence of the drone the testes reach a length of 5 millimeters and are now bean-shaped bodies of a creamy yellow color. After emergence they shrink, become flattened, and acquire a greenish yellow color. When the drone becomes functionally mature, 12 days after emergence, the testes have become so reduced that they are less than one-third their maximum length. The spermatozoa have now passed down into the seminal vesicles, and their migration is followed by the shrinkage of the testes.

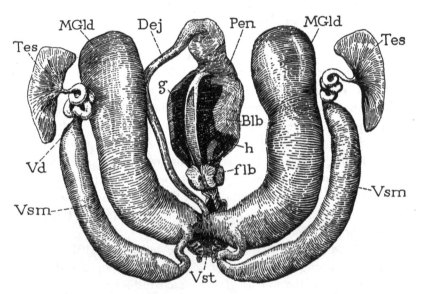

Fig. 101. Internal reproductive organs of the drone, somewhat spread out, dorsal view.

g, dorsal plate of penis bulb; h, lateral plate of penis bulb. For explanation of abbreviations see pages 314–315.

The undifferentiated male germ cells, known as the *primary spermatogonia,* are lodged in the upper ends of the tubules of the developing testes. In many insects it has been observed that they are arranged about a large apical cell from which they appear to derive nourishment. Each primary spermatogonium divides into a group of secondary spermatogonia, and the latter become encapsulated in a thin-walled cyst, within which they will develop into *spermatocytes.* The origin of the spermatogonia in the honey bee has not been observed, but Meves (1907) describes their later development. In the larva each cyst (fig. 102 B, *Cst*) is filled with pyramidal spermatogonia (B, C) that have their bases against the cyst wall and their inner ends convergent at the center of the cyst, where they are connected by granular extrusions (E, *a*). Scattered among the cysts are irregular cells (B, *FCl*) that eventually will enclose the cells of each cyst in a follicle (D). During the course of development the spermatogonia undergo divisions until they are ready for their transformation into spermatozoa. Maturation

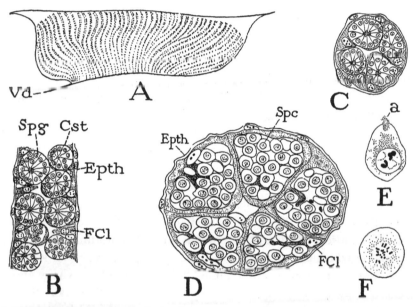

Fig. 102. The testes and developing sperm cells (A, from Nelson, 1924; B to F from Meves, 1907).

A, testis of larva, lateral, showing testicular tubules (length 3.75 mm.). B, lengthwise section of sperm tubule of adult drone, showing outer epithelium (*Epth*), sperm cysts (*Cst*), spermatogonia (*Spg*), and follicle cells (*FCl*). C, cross section of same. D, same, later stage, spermatogonia now developed into spermatocytes (*Spc*), cysts enclosed in follicles formed of the follicle cells. E, spermatogonium with extruded body (*a*). F, section of spermatogonium in division, showing 16 chromosomes.

now takes place (for details see Manning, 1949a), and the spermatocytes are converted into immature spermatozoa known as *spermatids*, which still have the form of ordinary cells. The spermatids are finally transformed into mature spermatozoa having the form of long threads with a swelling, or head, at one end containing the nucleus. All these changes take place in the tubules of the testis. According to Bishop (1920a), the final development of the spermatozoa is accomplished during the four days previous to the emergence of the drone from the comb cell. The spermatozoa then go down the vasa deferentia into the seminal vesicles, which are filled with sperm three days after the emergence of the drone.

Some of the older ideas concerning the number of chromosomes in the germ cells, the maturation of the egg, and sex determination in the honey bee must now be revised in the light of more recent work by Manning (1949a, 1949b, 1949c), Kerr (1951), and others.

The Vasa Deferentia and Vesiculae Seminales— Each vas deferens is differentiated into three parts. The duct leaves the testis as a short, tightly coiled tube (fig. 101, *Vd*), followed by a long, thick, sausage-shaped enlargement, which is the vesicula seminalis (*Vsm*), and the vesicula terminates in a short duct that opens into the dorsal wall of the base of the mucus gland (*MGld*) of the same side. The cells in the tubular parts of the vas deferens are said by Bishop (1920a) to be cubical in form, while those of the vesicula are long and narrow. Secretion in the vesiculae is discharged by constriction of the inner ends of the cells accompanied by dissolution of the cell substance cast off. The epithelium is covered by an inner layer of circular muscle fibers and an outer layer of longitudinal fibers, external to which is a membranous tunic. The spermatozoa on leaving the testicular tubules are retained in the vesiculae, where their heads become embedded in the epithelium with their tails filling the axial lumen. Here the spermatozoa attain their final form.

The Mucus Glands— The two great, curved saclike mucus glands of the drone bee (fig. 101, *MGld*) represent the male accessory genital glands of other insects, which are outgrowths of the posterior ends of the vasa deferentia. Each gland has an epithelial wall of glandular cells and a muscular sheath of outer longitudinal fibers and inner circular fibers. According to Bishop (1920a), a third innermost layer of longitudinal fibers in three bundles extends from the base of the gland more than halfway to the anterior end. The musculature is strongest around the mouth of the ejaculatory duct. The mucus glands are said by Bishop to increase in size during the first nine days after emergence of the drone. "The secretion changes in character from fluid to viscous, and acquires increasingly the property of immediately coagulating to a tough, cheesy or doughy mass." It is slightly alkaline and in its property of coagulating on

294

contact with air or water differs from the spermatic fluid secreted in the seminal vesicles.

The Ductus Ejaculatorius— The ejaculatory duct is a long slender tube (fig. 101, *Dej*) going from the anterior end of the inverted penis posteriorly to the united ends of the mucus glands. The wall of the duct is entirely nonmuscular. Where the duct comes into contact with the junction of the mucus glands, it divides into two short branches that penetrate the muscular sheaths of the glands opposite the openings of the vasa deferentia. Bishop (1920a) contends that there are no openings from the glands into the duct prior to the time of copulation and that the apertures are then produced as ruptures of the thin separating walls caused by violent contraction of the surrounding muscles of the glands. This contraction at the same time brings the mouths of the vasa deferentia into contact with the newly formed openings into the duct and thus allows the discharge of the spermatozoa into the penis. Following the spermatozoa comes a discharge of mucus from the glands. According to Bishop's findings, there should be no sperm in the penis before mating. On the other hand, Zander (1911) says that all the spermatozoa are forced through the ejaculatory duct into the bulb of the penis during the first days after the emergence of the drone, and that in the bulb, with mucus from the mucus glands, they form a sperm packet held ready for discharge. The writer has found in most drones examined a mass of spermatozoa and granular matter in the bulb of the penis as described by Zander. The matter will be further discussed in connection with the subject of *mating* (p. 306).

The Intromittent Organ— The functional penis of the male bee is a highly developed endophallic structure, represented in most other insects by the eversible membranous lining of the ectophallic organ known as the *aedeagus*. The ectophallic parts of the genitalia in the drone bee, as explained in chapter VIII, are a pair of elongate lobes, or penis valves (fig. 103 E, *pv*), closing over the phallotreme (*Phtr*) and a pair of small, setigerous lateral plates (*lp*) overlapping the bases of the penis valves. In most of the higher insects the penis valves are united to form the tubular aedeagus, and the lateral plates, or parameres, form a pair of copulatory claspers. The functional penis is usually the eversible membranous endophallic tube of the aedeagus. The complex intromittent organ of the drone bee, therefore, is entirely a greatly elaborated endophallus (fig. 103 D, *Enph*), while the ectophallic parts are much reduced or retain a very simple primitive structure.

The uneverted penis lies in the ventral part of the abdomen and extends forward to the posterior end of segment III. The organ is differentiated into three major parts (fig. 103 A, D). First, opening from the phallotreme (*Phtr*), is a wide, thin-walled end chamber, or *vestibulum* (*Vst*), from the sides of which project a pair of large, tapering, pouchlike *cornua* (*crn*), the "pneumophyses" of earlier writers. Beyond the vestibulum the penis narrows to a slender neck, or *cervix* (*Cer*), which is ordinarily compressed into an irregular shape (D) but when stretched out (A) is seen to be a fairly wide tube. On its dorsal surface it bears a large, doubly pinnate *fimbriated lobe* (*flb*), and along its ventral wall is a rounded fold (*e*) crossed by a row of internal V-shaped sclerites conspicuous externally by their dark color. The end of the cervix supports the large terminal expansion of the penis known as the *bulb* (*Blb*), into which opens the ejaculatory duct (*Dej*).

The penis bulb varies in size and shape according to its contents (fig. 103 A, D, *Blb*). The walls are mostly thin and flexible, but on the posterior part of the dorsal side there are two large, elongate, posteriorally tapering plates (*g*) separated by a narrow median external groove. The plates are covered externally by the thin epithelium of the bulb wall (C, *g*), but beneath them is a deep, brownish, semitransparent thickening of the cuticular intima (*i*), which is extended as a pair of thin plates into the lateral walls of the bulb. The writer (1925) formerly described this tissue below the plates as a "gelatinous thickening," but it is clearly a solid part of the cuticle. Branching from the posterior part of each dorsal plate of the bulb is a small triangular lateral plate (A, D, *h*) with a spiny margin. The inner walls of the bulb are entirely smooth (B). The bulb lumen opens into the cervix by a triangular aperture (*k*) between the posterior ends of the lateral plates (*h*). If the bulb is distended as at D of the figure, it will be found to be filled with a mass of granular secretion and spermatozoa (C, *j*).

The membranous areas of the inner walls of the vestibulum and cervix of the penis are everywhere thickly covered with minute spicules directed posteriorly. In the dorsal wall of the cervix is a thick, elongate triangular plate (fig. 103 A, *f*) with a grooved inner

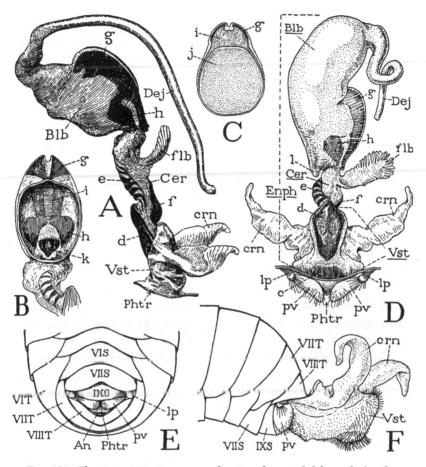

Fig. 103. The intromittent organ and external genital lobes of the drone.
A, the inverted penis, left side, bulb empty. B, interior of penis bulb, seen from below. C, cross section of bulb through dorsal plates (g) containing mass of seminal fluid and spermatozoa. D, penis and external genital lobes, dorsal, penis bulb distended. E, end of male abdomen, ventral. F, end of abdomen with penis partly everted.

c, posterior dark area of ventral wall of vestibulum; d, ventral plate of vestibulum; e, row of ventral V-shaped sclerites of cervix; f, dorsal plate of cervix; g, dorsal plate of bulb; h, lateral plate of bulb; i, sclerotic inner wall of bulb beneath dorsal plates; j, coagulated content of bulb; k, aperture from bulb into cervix; l, pocket of bulb wall. For explanation of abbreviations see pages 314–315.

297

surface. The V-shaped sclerites (*e*) in the ventral wall have bare inner surfaces, but the spaces between them are bristling with short hairs. The ventral wall of the vestibulum contains a large, dark shield-shaped area (A, D, *d*) the inner surface of which is densely hairy, and behind it is a smaller darkened area (D, *c*) covered internally with short hairs.

THE FEMALE ORGANS

The reproductive system of the female insect includes the *ovaries*, the paired *lateral oviducts*, and a median *common oviduct*. In the more generalized insects the common oviduct opens into a pocket at the posterior end of the body, known as the female *genital chamber*, between the eighth and the ninth abdominal sterna. In the honey-bee queen the genital chamber is differentiated into a *bursa copulatrix* opening at the base of the sting and a so-called *vagina* that receives the common oviduct in its anterior end. A sperm receptacle, or *spermatheca*, opens by a short duct from the dorsal wall of the vagina. Although the sting of the bee is the ovipositor of other insects, it is not used by the queen for egg-laying purposes in the manner of an ovipositor and hence has been fully described as a stinging organ in the chapter on the abdomen.

The Ovaries— The fully developed ovaries of the queen bee (fig. 104, *Ov*) consist of two huge, pear-shaped masses of egg tubules, or *ovarioles* (*Ovl*). It is generally estimated that each ovary contains 160 to 180 tubules. The ovarioles arise from the anterior ends of the lateral oviducts (*Odl*) and taper to slender threads, which finally in each ovary unite in a single suspensory strand attached beneath the ventral wall of the heart. An ovariole looks like a string of beads increasing in size toward the oviduct. The bead-like swellings are due to the contained eggs in successive stages of development together with accompanying nutritive cells. The outer wall of each tubule is a thin cellular epithelium (fig. 105 A, B, D, G, *Epth*).

The extreme upper threadlike part of an ovariole contains only a multinucleate protoplasmic mass in which cell boundaries are not evident, but a little farther down in the tubule (fig. 105 A) cell boundaries are distinct and the core of the tubule is here a solid row of small cells. These cells are the *primary oogonia* (*Oog*). Still farther along in the tubule (B) the oogonia break up into a mass

298

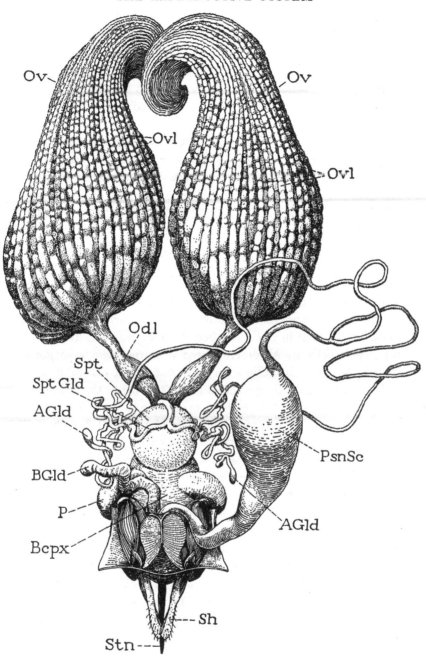

Fig. 104. Reproductive organs and the sting of the queen, dorsal view. For explanation of abbreviations see pages 314–315.

299

of irregularly distributed polygonal cells, and the diameter of the tubule has increased to accommodate the added bulk of its contents. It is now to be seen also that, in addition to the oogonia, other smaller cells (*FCl*) are present lying against the epithelial wall.

At a lower level (fig. 105 C) the oogonia themselves have differentiated into large cells (*Ooc*), distributed irregularly along the axis of the ovariole and more numerous interspersed small cells (*NCl*), while the superficial cells (*FCl*) now form a definite outer layer just within the epithelium. The large axial cells are the future egg cells and are known as the *oocytes*. The intervening small cells are commonly called "nurse cells," but they will be consumed as food by the growing oocytes and are hence better termed *food cells*, or *trophocytes*. The superficial cells will later form *follicles* about the oocytes and their food cells. The beginning of the follicular structure of the ovariole is seen at D of the figure; each oocyte and the group of trophocytes above it are here surrounded by a layer of the follicle cells (*FCl*). The follicle cells are commonly regarded as derivatives from the epithelium of the ovariole, but Manning (1949b) says that in the honey bee the oocytes, the nurse cells, and the follicle cells are all differentiated from the primary oogonia. If such is the case, the follicle and its contained cells might be said to be a preliminary reproductive organism within the ovariole.

Eventually the follicle cells form a definite capsule about each oocyte and its accompanying food cells (fig. 105 E), but the follicle cells become particularly large and compactly arranged around the oocyte. At this stage also the follicle cells begin to form a circular fold between the oocyte and the trophocytes that divides the follicle into an egg chamber and a nutritive chamber. The fold later becomes deeper (F) until there is only a central opening between the two chambers, into which the upper end of the oocyte protrudes as a blunt plug against the adjoining food cells. The trophocytes grow rapidly by the absorption of nutritive material from the blood through the follicle walls until the nutritive chambers become much larger than the egg chambers (G). In the lower ends of the ovarioles, however, the relative size of the two chambers becomes reversed owing to the growth of the oocytes at the expense of the trophocytes. Finally, when the trophocytes are fully consumed by the oocytes, the last follicles of the ovarioles contain only the oocytes, which now are fully grown ova. The maturation of the honey-bee egg is described by Manning (1949c).

It is probable that each oocyte and its food cells are descendants of a single primary oogonium. According to Paulcke (1900), each

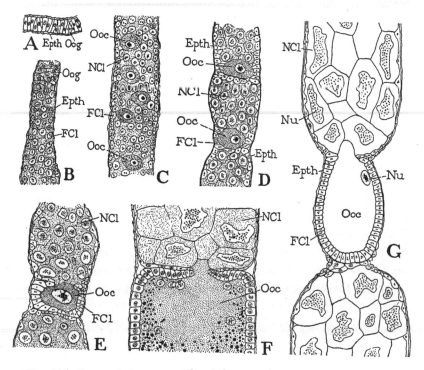

Fig. 105. Oogenesis in an ovariole of the queen's ovary.

A, upper end of ovariole containing undifferentiated oogonia. B, section of tubule farther down, oogonia multiplied and surrounded by smaller follicle cells. C, still farther down, oogonia differentiated into oocytes (*Ooc*) and trophocytes, or nurse cells (*NCl*). D, formation of follicles by the follicle cells enclosing group of trophocytes above and an oocyte below. E, later stage of same, follicle partly divided into nutritive chamber and egg chamber. F, upper end of an oocyte nearing maturity and adjoining trophocytes. G, diagram of an egg chamber and two nutritive chambers.

For explanation of abbreviations see pages 314–315.

oogonium of the queen bee, by two consecutive divisions, forms first four cells. One of these cells becomes an oocyte; the other three, by four further divisions, produce 48 trophocytes, the approximate number in each nutritive chamber of the ovariole. In the end, the 49 cells are again united in the ovum, the substance of the trophocytes being absorbed by the oocyte to form the yolk of the egg, and the yolk is finally converted into tissues of the embryo. Though 49 cells may thus contribute to the growth of the embryo, the hereditary factors, or genes, are carried in the chromosomes of the original oocyte only. In the lowermost chamber of the ovariole the egg is covered with a thin shell, or *chorion*, formed of a secretion from the follicle cells, and is now ready to be discharged into the oviduct.

The egg of the bee is cylindrical, rounded at each end, curved in its long axis, the convex side being the ventral side of the future embryo. It is pearly white, and, according to Nelson (1915), it varies in length from about 1.53 millimeters to 1.63 millimeters. The eggs of most insects have an aperture, the *micropyle*, or several apertures in the anterior end of the chorion for the entrance of spermatozoa. At the anterior end of the bee's egg, Nelson says, there is a distinct micropylar area, but he was unable to demonstrate the presence of actual perforations in it. As each egg matures and passes into the oviduct, the cells of the deserted follicle degenerate and are absorbed, or their debris is thrown out into the duct. The egg next above then descends in its follicle, and new oocytes are generated in the upper end of the ovariole. Throughout the life of the queen eggs are thus discharged as others mature, and it has been estimated that a queen in the prime of her reproductive vigor may produce as many as 3,000 eggs in a single day. The usual number, however, is probably closer to 1,500 eggs a day, and even at this rate the queen produces more than her own body weight in eggs every 24 hours. Her yearly output has been estimated at 200,000, but she does not maintain this rate throughout the four years of her average life.

The Oviducts— The paired lateral oviducts (fig. 104, *Odl*) converge posteriorly from the ovaries to join a short, common oviduct (fig. 106 E, *Odc*). Each lateral duct has a calyxlike expansion against the base of the corresponding ovary, into which the ovarioles open, and is somewhat enlarged in its posterior part. The duct walls are thrown into lengthwise folds that allow of much expansion. The lateral ducts are lined with a thin cuticular intima bearing backwardly directed hairs and are ensheathed in fine longitudinal muscle fibers. The short, wide common oviduct has strongly muscular walls, and its ventral wall contains a deep median channel. It opens immediately into the much wider saclike part of the genital exit passage commonly called the vagina (*Vag*).

In most insects the lateral oviducts are the primitive mesodermal outlets of the ovaries, which discharge into an ectodermal common duct, but in many cases the common duct branches and partly or entirely replaces the mesodermal ducts. Thus, in the honey bee it has been shown by Löschel (1916) and by Meier (1916) that the paired oviducts are derived from the ectoderm in both the queen and the worker. Consequently these ducts in the bee have a cuticular intima.

The Bursa Copulatrix and the Vagina— In the female bee both segment VIII and segment IX of the abdomen are drawn into segment VII and form the chamber that contains the sting (fig. 59). Segment VIII is a complete annulus in the inner end of the sting chamber, but its walls are membranous except for the two small spiracle-bearing plates (*Lsp*), which are remnants of the tergum. The genital chamber of the queen bee opens from the anterior wall of the sting chamber at the base of the sting. Its outer part is the bursa copulatrix (fig. 104, *Bcpx*), the inner part the vagina (fig. 106 E, *Vag*). The common oviduct (*Odc*) opens into the anterior end of the vaginal region and the spermathecal duct (*SptDct*) through the dorsal wall of the latter. The anatomical relations of the external genital passage in the queen bee, therefore, are the same as those of the female genital chamber in other insects.

The bursa copulatrix is a wide-open pocket within which the somewhat bilobed base of the sting is freely lodged. The interior of the bursa is seen at E of figure 106 inside the cut margin (*x*) of its wall. In its inner wall are three conspicuous openings: one median (*VO*) is the entrance to the vagina; the two larger lateral openings (*PO*) lead into a pair of large pouches (*P*) projecting from the sides of the bursa. On the dorsal wall of each pouch is attached a strong muscle (*185*) from the basal margin of sternum VII.

The vagina is an oval sac, but it varies in form and apparent size according to its distension. The ventral wall is inflected to form a high, transverse, valvelike fold projecting into the lumen below the orifice of the spermathecal duct. The fold has a strong musculature from sternum VII, described by Laidlaw (1944) and by F. Ruttner (1956). Two pairs of slender muscles (fig. 106 E, *193, 194*) from the spiracle plates of segment VIII are attached on the dorsal wall

303

of the vagina. Some writers have described the short, common oviduct of the queen (*Odc*) as a narrowed anterior part of the vagina, but it clearly represents the median oviduct of other insects

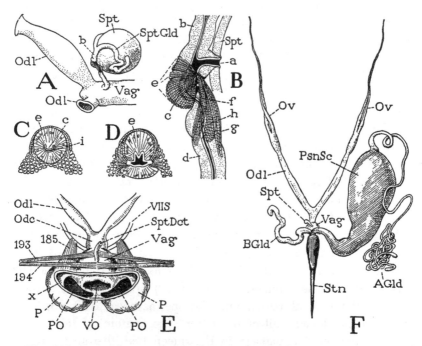

Fig. 106. Spermatheca and genital exit passages of a queen and ovaries of a worker (A, B, C, D from Bresslau, 1906).

A, anterior end of vagina with right lateral oviduct and spermatheca. B, upper end of spermathecal duct (*d*) showing muscles and opening (*a*) into spermatheca. C, cross section through upper part of spermathecal duct. D, section of duct through muscle attachments. E, genital apertures and passages of the queen, dorsal. F, ovaries, sting, and sting glands of a worker, dorsal.

a, opening of spermathecal duct into spermatheca; *b,* duct of spermathecal gland; *c,* anterior fold of wall of spermathecal duct; *d,* spermathecal duct; *e,* semicircular muscles of duct; *f, g, h,* longitudinal muscles on duct wall; *i,* lumen of duct; *x,* cut margin of wall of bursa couplatrix. For explanation of abbreviations see pages 314–315.

which opens into the genital chamber, the last being the functional exit passage for the eggs in the bee.

The Spermatheca— The spermatheca of the queen bee is a fairly large, globular sac (figs. 104, 106 A, *Spt*) lying over the vagina, with which it is connected by a short duct. The upper end of the spermathecal duct is joined by the duct (fig. 106 A, *b*) of a pair of tubular spermathecal glands (*SptGld*) closely applied to the surface of the spermatheca.

Inasmuch as the fertile queen is able to control the insemination of the eggs in the vagina, much interest attaches to a special structure of the spermathecal duct described by Bresslau (1905) as a mechanism for sperm discharge. The upper part of the duct, just below its entrance into the sperm sac (fig. 106 B, *a*), makes a sigmoid flexure, of which the dorsal loop is more strongly convex than the ventral. Just above the upper bend the duct from the spermathecal gland (*b*) opens into the spermathecal duct (*d*). In the upper bend of the latter a fold (*c*) of the anterior wall of the duct projects into the lumen, which is thus reduced to a narrow crescentic slit (C, *i*). A layer of semicircular muscle fibers (B, *e*) forms a caplike covering over the convexity of the upper loop of the duct, the fibers being attached on lateral extensions of the thick cuticular intima (D). At the same points are attached two lateral groups of longitudinal fibers (B, *f*, *g*) that arise ventrally on the lower straight part of the duct. A third muscle (*h*) on the posterior wall of the duct is attached dorsally at the end of the lower bend. The probable role of this mechanism in the discharge of sperm from the spermatheca will be discussed later.

The Reproductive Organs of a Worker— Inasmuch as the worker bee is a female, she is equipped with reproductive organs, though only on special occasions do they become functional. The ovaries (fig. 106 F, *Ov*) are greatly reduced, each, according to Meier (1916), consisting of 2 to 12 slender ovarioles. The lateral oviducts (*Odl*) are relatively long tubes converging to a short, common oviduct, which opens into a small vaginal pouch (*Vag*) at the base of the sting. Projecting from the dorsal wall of the vagina is a vestigial spermatheca (*Spt*).

It is highly interesting to note from the studies of Meier (1916) on the development of the reproductive organs in the worker bee that in the early larval stages the ovaries of the worker are essentially the same as those of the queen. In a two-day-old worker larva Meier found 62 ovarioles in one ovary and 72 in the other; in a three-day larva there were 72 and 89 ovarioles. At this age, therefore, the ovaries of both the worker and the queen are in about the same stage of development. By the fifth day the worker ovary attains its maximal development with an average of about 130

305

ovarioles, but here development ceases. At the beginning of the pupal stage 110 or 115 ovarioles may be present, but during metamorphosis most of the ovarioles degenerate, leaving only 2 to 12 much-shrunken tubules in each ovary of the mature worker. The worker larva is thus potentially a functional female, and the state of the ovaries in the adult worker results from both inhibition and retrogressive development.

The reduced worker ovaries, however, remain functionally competent, and when a bee colony becomes queenless the ovaries of many of the workers renew their development and are able to produce normal eggs (Leuenberger, 1927). The nature of the stimulus that induces functional redevelopment in the ovaries is not exactly known (see review of "laying workers" by Ribbands, 1953). Since worker eggs are not fertilized, they ordinarily produce only drones, so it is not clear what advantage the colony derives from them. However, rare cases are known in which worker eggs and also some eggs of unfertilized queens develop into perfect females (Mackensen, 1943), probably by failure of the reduction division in maturation, leaving them in the diploid female condition, as suggested by Manley (1948). Although it is said that drones are sometimes seen "mated" with workers, it is highly unlikely that insemination takes place.

The behavior of the egg-laying worker, as described by Örösi-Pál (1932), would indicate that the instincts of the worker for the laying of eggs are much less developed than are those of the queen. She first inspects a cell in the queen's manner by inserting her head and the anterior part of her thorax, and then inserts her abdomen to deposit the egg, but she is as likely to stick the egg anywhere on the sides of the cell as at the bottom. Then too she often puts an egg into a cell already containing an egg or even a larva. She spends much time wandering over the comb, going from one cell to another without laying. This habit Örösi-Pál suggests is probably due to the smallness of her ovaries, which cannot produce at once as many eggs as do those of the queen, so there are longer intervals between eggs entering the vagina.

MATING

Much literary sentiment has been wasted on the queen bee and her marriage flight. She mates on the wing, but she does not necessarily soar into the "azure heights" pursued by a swarm of ardent suitors, nor is she always true to the first drone with whom she mates. Him she leaves dead or dying on the field, and the very next day she may go out and mate with another, or even repeat her infidelity on subsequent occasions. That multiple mating is common practice with the honey-bee queen is now well attested by the work of a number of investigators, including Roberts (1944), S. and F. Ruttner (1954), Alber, Jordan, and F. and H. Ruttner (1955). According to the last authors, double matings are the most frequent, but three, four, and even five matings by a single queen have been observed. S. and F. Ruttner record an example of a queen that mated twice on the same day. The successfully mated queen returns to the hive with her marriage certificate in the form of a part of the male organ that inseminated her retained in her sting chamber. The other bees accept this "mating sign" as a guarantee of her maternal competence, and promptly proceed to remove it along with the mucus in her genital passage.

Inasmuch as the act of mating is performed very quickly while the bees are on the wing, exactly what takes place is difficult to observe and is still a subject of speculative difference of opinion, fully reviewed by Laidlaw (1944). In *The ABC and XYZ of Bee Culture* (A. I. and E. R. Root, 1917) are quoted accounts of mating as seen at close range by several observers. Some report that the drone clasps the queen from above, others that the two meet "face to face," i.e., the drone clinging to the underside of the queen (face to face would be head on with insects), and that the pair then flies off in a vertical position. Bishop (1920b) says, "The insects meet and clasp face to face; the female on being grasped allows the tip of the male's abdomen to enter the genito-anal vestibule; the drone, by explosive contraction of the abdominal walls, everts the organ (penis) into the female genital tract." Suddenly is heard, according to most observers, a loud explosive snap, after which the drone falls dead to the ground, usually bringing with him the queen, who quickly liberates herself and carries off in her sting chamber a detached part of the male organ.

There has been much discussion as to whether the entire penis, including the bulb, is everted during mating or whether only the uneverted bulb is injected into the queen. The question should be resolved by examining mated queens, but here again observations do not all agree. The writer's experience is not wide in this matter, but two recently mated queens received from Dr. E. Oertel of Baton Rouge, Louisiana, had each a mass of white material projecting from the sting chamber. An examination of the mass showed that it was the *uneverted* bulb of the penis greatly distended with coagulated mucus. On the undersurface were the dorsal and lateral plates of the bulb with the posterior prongs directed forward in the queen (fig. 107 C), and the lateral wings of the inner cuticle over the plates extended upward in the sides of the bulb. The penis bulb was, therefore, inverted and reversed in position in the queen, and S. and F. Ruttner (1954) say this is the usual position of the organ as found in the queen. A paper by Woyke (1955) on multiple mating in one nuptial flight has been received too late for discussion in the present text, but the author says that "almost all queens return with a noneverted bulb of the endophallus."

The inverted and reversed position of the penis bulb in the queen suggests that during mating the drone clung to the back of the queen, for in this position of the mated pair the bulb, by being merely inserted in the natural position, would be inverted in the sting chamber with its open end directed forward toward the mouth of the vagina. Zander's figure (1911, fig. 116, III) of the *Begattungszeichen* in the queen is correct, except for the minor error of showing the lateral plate of the bulb attached to the anterior end of the inverted dorsal plate. In copulation, Zander says (p. 126), the penis (*Begattungsschlauch*) is everted only as far as the bulb, and, according to his observations, only the bulb, which has been torn from the drone's body, is found in the queen. Zander does not explain how the neck of the bulb is severed from the rest of the organ nor how the sperm-filled bulb is discharged, but presumably the muscular contraction of the abdomen is the effective force.

The penis bulb in unmated drones, as found by the writer in drones taken from June to late summer, is filled with a mass of white granular material. In specimens preserved in alcohol this material forms a dense body that can be removed from the bulb like a seed from a pod (fig. 107 B). The body is what Zander (1911) described as a spermatophore (*Samenpatrone*), but it is not a true spermatophore. In freshly killed specimens the bulb content is a thick, white, minutely granular, somewhat viscous milky liquid containing myriads of writhing spermatozoa. The liquid does not coagulate on contact with the air, but it quickly dries. According to Zander, the spermatozoa are discharged into the penis bulb during the first few days after the emergence of the drone, who from now on is thus "loaded and ready to shoot." When the penis is only partly everted, as it usually is by artificial pressure on the abdomen, the sperm-filled bulb lies in the everted vestibulum. Some recent investigators deny the presence of sperm in the bulb before mating, but perhaps they examined drones that were too young; or, admitting that a mass of sperm is sometimes found in the bulb, they explain its presence as due to previous handling of the drones. If the bulb does not serve for the storage of sperm, we may ask what is the purpose of it, and why does not the drone bee have a simple, eversible endophallic tube such as that of many other male insects?

In the specimens of mated queens examined by the writer, in which the uneverted bulb of the penis was held in the sting chamber, the spermatic fluid containing the spermatozoa had already been discharged into the greatly distended oviducts, and the bulb itself was tensely filled with mucus. The mucus could have entered the bulb only by way of the ejaculatory duct. The duct, therefore, probably retains its connection with the bulb in the sting chamber until it is broken off by the efforts of the queen to free herself from the drone. Zander (1911) notes that a piece of the duct remains attached to the bulb. Perhaps it is the forcible discharge of the mucus into the bulb that drives the spermatic fluid through the genital passage into the oviducts; otherwise the filling of the oviducts from the open mouth of the bulb is difficult to understand.

That the complex genital organ of the drone bee is capable of an entire eversion can be demonstrated by compressing the abdomen of a live specimen, and some investigators contend that in normal mating it is the everted bulb of the penis that is thrust into the queen. Bishop (1920b), in discussing the extent of the penis eversion in copulation, records that, of three mated queens examined by him, in two the bulb was fully everted, in one apparently not.

The drawing given here of the everted penis (fig. 107 A) was made from a drone immediately after it was killed in a bottle with a very weak dose of carbon tetrachloride. The organ is more fully extended than usually seen and was tensely distended with air, the source of which was not determined. The distal part of the bulb is sharply bent upward and forward, with the dorsal plates (g) curved around the angle and the lateral plates (h) standing vertically just proximal to the bend. The dorsal plates are not exposed on the surface as they generally appear to be in illustrations but are inside

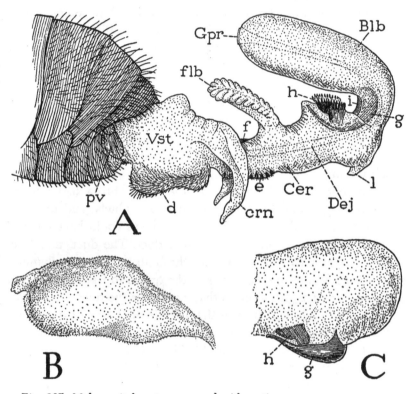

Fig. 107. Male genital parts concerned with mating.
A, the fully everted penis with gonopore (Gpr) at apex of bulb. B, body of coagulated seminal fluid and spermatozoa taken from the bulb of a drone preserved in alcohol. C, anterior end of uneverted penis bulb from sting chamber of a queen, showing the bulb plates inverted and reversed on undersurface; the plates are covered by a thin external membrane of the bulb wall.
Lettering as on figure 103.

the everted bulb, though plainly seen through its thin lateral walls. The smooth, hard surface (i) in the angle of the bulb is that of the thick cuticular layer that lies beneath the plates in the uneverted bulb (fig. 103 C, i). The position of the plates in the angle of the everted bulb would seem to make it quite impossible for the bulb

310

to turn downward. The curvature of the plates prevents anything but a dorsal flexure, and we may suspect that it is the function of the plates to cause and maintain the upward bend.

When the penis is fully everted (fig. 107 A), the ejaculatory duct (*Dej*) is drawn clear through the organ and the gonopore (*Gpr*) is brought to the apex of the bulb. It is contended by Shafer (1917) and by Bishop (1920a) that the sperm is ejaculated from the seminal vesicles at the instant of mating. Considering the fact that the ejaculatory duct is a slender nonmuscular tube, which in traversing the everted penis curves closely around the convex sides of the bulb plates (*g*), and furthermore that a great mass of sperm is driven through the genital passage of the queen with sufficient force to distend the lateral oviducts, the thin muscular walls of the vesiculae must be supposed to have extraordinary strength. Moreover, the process of insemination from the vesiculae would certainly be slow and time-consuming, whereas ejaculation takes place almost instantaneously with the mated bees and is accompanied by an explosive sound. The drone penis evidently is made for quick action, to expel a large quantity of seminal fluid and spermatozoa instantaneously and with great force. After the expulsion of the semen there takes place a discharge of mucus, presumably through the ejaculatory duct from the strongly muscular mucus glands. A mucus discharge, however, has not been observed by the writer in the artificially everted penis.

The eversion of the penis is produced by the contraction of the abdomen, and this function of the abdomen probably accounts for the great size of the abdominal muscles in the drone (fig. 54 B) as compared with the corresponding muscles in the worker (A). In the uneverted condition of the penis there is a sharp constriction between the bulb and the cervix (fig. 103 A, D), and there is only a relatively small aperture between the two parts (B, *k*), which effectively holds back the spermatic fluid in the bulb. The lips of the fold, however, are evidently completely distensible, since in the protracted penis (fig. 107 A) the lumen of the cervix is widely open into that of the bulb. Ventral to the aperture a small pocket projects from the end of the bulb (figs. 103 D, 107 A, *l*). It seems remarkable that the large bulb with its hard plates can be everted into the narrow cervix. Shafer (1917), however, says that during eversion of the bulb the dorsal plates of the latter can be seen to make a "somersault" by swinging downward and posteriorly on their prongs as on a pivot, which they are able to do because of the thin ventral wall of the bulb.

If it is true that in some cases the insemination of the queen is accomplished by ejaculation of the seminal fluid and mucus through the ejaculatory duct within the everted bulb, it becomes a question as to what is the position of the mated pair relative to each other. The dorsal and forward curvature of the bulb in the fully everted condition of the penis (fig. 107 A) would indicate that the queen rides on the back of the drone, but this position in mating has not been observed. On the other hand, if the drone either clasps the queen from above or clings upside down beneath her, the bulb would have to turn ventrally to enter the sting chamber. Shafer (1917) contends that the drone organ "must" turn downward and forward to accomplish its purpose. Yet the penis is entirely non-muscular, and, as above noted, the position of the plates in the angle of the everted bulb appears to make a downward flexure of the bulb quite impossible. Further discussion of the subject will be useless until more facts are known, but it may be observed that if the sperm and mucus are ejaculated from the bulb, the bulb in the queen should be collapsed within the mucus instead of being filled with it. Finally, it is possible that the bulb may sometimes be entirely withdrawn, since Fyg (1952) reports that in his experience he found only mucus in the queen.

Whatever the facts may be as to how the queen becomes inseminated, all writers agree that the spermatozoa are first stored in the distended lateral oviducts. The prevalent opinion has been that the spermatozoa make their way back to the spermatheca by their own activity, probably activated by chemotactic stimulation. F. Ruttner (1956), however, has shown that the transfer is made otherwise; it begins as soon as the remains of the penis and the mucus are removed from the queen. Now, at intervals of one or two minutes, the queen stands with her abdomen decurved while a trembling movement runs through her body as if caused by an active muscular compression. The spermatozoa, which have of themselves taken a parallel arrangement, are thus forced out of the oviducts and into the anterior end of the vagina. Here their farther progress is blocked by the elevation of the vaginal valve, and the sperm are then forced up through the spermathecal duct into the spermatheca. The time consumed in the filling of the spermatheca varies with different individuals. Five hours after mating, Ruttner says, there may still be a considerable amount of sperm in the oviducts, but after 24 hours the ducts are always free of sperm and mucus.

The supply of spermatozoa received by the queen at one mating is generally sufficient to last two or three or perhaps four years of the queen's life. If the quantity of sperm should be less than normal, or if for any other reason the queen has an urge for a second marriage, she is now free to fly out for another mating if she is so inclined. When a queen begins laying eggs, however, she seldom if ever makes any further mating flights.

INSEMINATION OF THE EGGS

It is well known that the egg-laying queen deposits fertilized eggs in the smaller cells of the comb and unfertilized eggs in the larger cells, except as she sometimes seems to be confused where the two sets of cells adjoin and may put a fertilized egg in a large cell. In general, therefore, eggs in small cells give rise to workers and those in large cells to drones. The proportional number of workers and drones in the colony is determined by the workers that construct the cells. Aside from these few facts, we are ignorant as to why the workers make cells of two sizes, and it has been a long-standing mystery with beekeepers as to how the queen regulates insemination of an egg according to the size of the cell. Moreover, the size of the cell is not an invariable guide because queen cells are even larger than drone cells and yet they receive fertilized eggs.

The insemination of an egg takes place in the vagina as the egg passes the orifice of the spermathecal duct. The discharge of the spermatozoa was attributed by Bresslau (1905) and by Adam (1913) entirely to the action of the muscular "sperm pump" of the spermathecal duct (fig. 106 B), though these two writers explained the action of the pump in different ways. Adam suggested that the valvular fold on the floor of the vagina, when erected, serves to hold a passing egg that is to be fertilized against the orifice of the spermathecal duct to receive the sperm. If the egg is not to be fertilized, the fold relaxes and activity of the pump ceases. Neither Bresslau nor Adam, however, attempted to explain how the mechanism of insemination is controlled by the queen, but it was formerly assumed that when the queen inserts her abdomen into a cell the closer contact with the walls of the smaller cells gives the stimulus for action by the sperm pump.

Before the laying queen deposits an egg in any cell, she first briefly inspects the cell by thrusting her head and antennae into it. Flanders (1950), therefore, suggests that the stimulus for sperm discharge comes from the antennae, and, moreover, he contends that primarily it influences secretion in the spermathecal gland. The gland secretion, being alkaline in contrast to the weakly acid fluid in the spermatheca, activates the spermatozoa, which move out from the spermatheca into the upper part of the duct. Here the gland secretion accumulates and the increasing pressure tends to keep the valve of the duct closed. Then as an egg passes into the vagina, it causes the valve to open for an instant and to emit a few spermatozoa in a minute quantity of secretion. Flanders thus attributes the primary stimulus for insemination to the antennae instead of to the abdomen and assigns only a final role in the action of the insemination mechanism to the sperm pump. As presented it does not account for the laying of fertilized eggs in queen cells, but in private correspondence Flanders points out that the queen cells are constructed singly in proximity to worker cells and that in first going from a worker cell to a larger cell the queen commonly deposits in the latter a fertilized egg, owing to some lag in the response of the spermathecal mechanism.

Explanation of Abbreviations on Figures 101–107

AGld, poison gland of sting.
An, anus.

Bcpx, bursa copulatrix.
BGld, accessory gland of sting.
Blb, bulb of penis.

Cer, cervix of penis.
crn, cornua of penis.
Cst, sperm cyst.

Dej, ductus ejaculatorius.

Enph, endophallus (inverted penis).

Epth, epithelium.

fbl, fimbriated lobe of penis.
FCl, follicle cells.

Gpr, gonopore.

lp, lamina parameralis.

MGld, mucus gland.

NCl, nutritive cells of ovary (nurse cells, trophocytes).

Odc, oviductus communis.

314

Ooc, oocyte.
Oog, oogonium.
Ov, ovary.
Ovl, ovariole.

P, lateral pouch of bursa copulatrix.
Pen, penis (endophallus).
Phtr, phallotreme (aperture of inverted penis).
PO, opening of bursal pouch in bursa copulatrix.
pv, penis valve (mesomere).

Sh, sheath lobes of sting.

Spc, spermatocyte.
Spg, spermatogonium.
Spt, spermatheca.
SptGld, spermathecal gland.
Stn, shaft of sting.

Tes, testis.

Vag, vagina.
Vd, vas deferens.
VO, vaginal opening in bursa copulatrix.
Vsm, vesicula seminalis.
Vst, vestibulum of inverted penis.

REFERENCES

Adam, A. 1913. Bau und Mechanismus des Receptaculum seminis bei den Bienen, Wespen und Ameisen. Zool. Jahrb., Anat., 35: 1–74, 3 pls.

Alber, M., Jordan, R., and Ruttner, F. and H. 1955. Von der Paarung der Honigbiene. Zeitscher. Bienenforschung, 3: 1–28, 21 figs.

Anglas, J. 1901. Observationes sur les métamorphoses internes de la guêpe et de l'abeille. Bull. Sci. France et Belgique, 34: 363–473, 5 pls.

Armbruster, L. 1931. Nahrung, Verdauung und Stoffwechsel der Bienen. In Mangold, Handbuch der Ernährung und des Stoffwechsels der Landwirtschaftlichen Nutztiere, 3: 478–563, 42 figs. Berlin.

Autrum, H., and Stumpf, H. 1950. Das Bienenauge als Analysator für polarisiertes Licht. Zeitschr. Naturforsch., 5b(2): 116–122, 3 figs.

Bailey, L. 1952. The action of the proventriculus of the worker honeybee, *Apis mellifera* L. Journ. Exp. Biol., 29(2): 310–327, 7 text figs., 1 pl.

Bailey, L. 1954. The respiratory currents in the tracheal system of the adult honey-bee. Journ. Exp. Biol., 31: 589–593, 1 fig.

Beams, H. W., and King, R. L. 1933. The intracellular canaliculi of the pharyngeal glands of the honeybee. Biol. Bull., 64: 309–314, 4 figs.

Beling, Ingeborg. 1931. Beobachtungen über das Pollensammeln der Honigbiene. Arch. Bienenk., 12: 76–83.

Berlese, A. 1901. Osservazione su fenomeni che avvengono durante la ninfosi degli insetti metabolici. I. Tessuto adiposo. Rivista Patologia Vegetale, 8: 1–155, 42 text figs., 6 pls.

Berlese, A. 1909. Gli Insetti, vol. I. Milan.

Bertholf, L. M. 1925. The moults of the honeybee. Journ. Econ. Ent., 18: 380–384.

Betts, Annie D. 1923. Practical Bee Anatomy, 88 pp., 12 pls. The Apis Club, Benson, England.

Bigelow, R. S. 1954. Morphology of the face in the Hymenoptera. Canadian Journ. Zool., 32: 378–392, 21 figs.

REFERENCES

Bishop, G. H. 1920a. Fertilization in the honey-bee. I. The male sexual organs: Their histological structure and physiological functioning. Journ. Exp. Zool., 31: 225–265, 3 text figs., 3 pls.

Bishop, G. H. 1920b. Fertilization in the honey-bee. II. Disposal of the sexual fluids in the organs of the female. Journ. Exp. Zool., 31: 267–286, 2 figs.

Bishop, G. H. 1922. Cell metabolism in the insect fat-body. I. Cytological changes accompanying growth and histolysis of the fat body of *Apis mellifica*. Journ. Morph., 36: 567–600, 3 pls.

Bishop, G. H. 1923a. Cell metabolism in the insect fat-body. II. A functional interpretation of the changes in structure in the fat cells of the honeybee. Journ. Morph., 37: 533–553.

Bishop, G. H. 1923b. Body fluid of the honeybee larva. Journ. Biol. Chem., 58: 543–565.

Bragg, W. 1933. The Universe of Light, 283 pp., 110 text figs., 26 pls. New York.

Bresslau, E. 1905. Der Samenblasengang der Bienenkönigen. Zool. Anz., 29: 299–323, 7 figs.

Brunnich, K. 1922. Zum Atemmechanismus bei den Insekten, speziell bei den Bienen. Arch. Bienenk., 4: 157–160.

Bugnion, E. 1928. Les Glandes salivaires de l'abeille et des apiaires en général, 64 pp., 29 figs. Librairie de vulgarisation apicole, Montfavet, Vaucluse.

Butler, C. G. 1954a. The method and importance of the recognition by a colony of honeybees (*A. Mellifera*) of the presence of its queen. Trans. R. Ent. Soc. London, 105, 11–29, 1 fig.

Butler, C. G. 1954b. The World of the Honeybee, 226 pp., 40 pls. New York.

Butler, C. G. 1955. The role of "queen substance" in the social organization of a honeybee community. American Bee Journ., 95(7): 275–279, 3 figs.

Butt, F. H. 1934. The origin of the peritrophic membrane in *Sciara* and the honey bee. Psyche, 41: 51–56, 1 pl.

Cajal, S. R., and Sánchez, D. 1921. Sobre la estructura de los centros ópticos de los insectos. Revista Chilena Hist. Nat. Santiago, 25: 1–18, 2 pls.

Campbell, F. L. 1929. The detection and estimation of insect chitin; and the irrelation of chitinization to hardness and pigmentation of the American cockroach, *Periplaneta americana*. Ann. Ent. Soc. America, 22: 401–426.

Carlet, G. 1890. Mémoir sur le venin et l'aiguillon de l'abeille. Ann. Sci. Nat., Zool., ser. 7, 9: 1–17, 1 pl.

317

REFERENCES

Casteel, D. B. 1912a. The manipulation of the wax scales of the honey-bee. U.S. Dept. Agric., Bur. Ent., Circular No. 161: 1–13, 7 figs.

Casteel, D. B. 1912b. The behavior of the honeybee in pollen collecting. U.S. Dept. Agric., Bur. Ent., Bull. 121, 36 pp., 9 figs.

Chadwick, L. E. 1955. Molting of roaches without prothoracic glands. Science, 121: 435.

Chapman, J. A., and Craig, R. 1953. An electrophysiological approach to the study of chemical sensory reception in certain insects. Canadian Ent., 85: 182–189, 11 figs.

Child, C. M. 1894. Ein bisher wenig beachtetes antennales Sinnesorgane der Insekten, mit besonderer Berücksichtigen der Culiciden und Chironomiden. Zeitschr. wiss. Zool., 58: 475–528, 2 pls.

Day, M. F., and Waterhouse, D. F. 1953. Functions of the alimentary system. In Roeder, Insect Physiology, Chap. 11, 298–310.

Debaisieux, P. 1938. Organes scolopidiaux des pattes d'insectes. II. La Cellule, 47: 77–126, 9 pls.

Dehn, Madeleine von. 1933. Untersuchungen über die Bildung der peritrophischen Membran bei den Insekten. Zeitschr. Zellforsch. mikr. Anat., 19: 79–105, 12 figs.

Dobrovsky, T. M. 1951. Postembryonic changes in the digestive tract of the worker honeybee (Apis mellifera L.). Cornell Univ. Agr. Exp. Stat., Mem. 301, 45 pp., 12 pls.

Dreyling, L. 1903. Über die wachsbereitenden Organe der Honigbiene. Zool. Anz., 26: 710–715, 2 figs.

Duncan, C. D. 1939. A contribution to the biology of North American vespine wasps. Stanford Univ. Pubs, Biol. Sci., 8(1): 272 pp., 54 pls.

DuPorte, E. M., and Bigelow, R. S. 1953. The clypeus and the epistomal suture in Hymenoptera. Canadian Journ. Zool., 31: 20–29, 21 figs.

Edwards, G. A., and Ruska, H. 1955. The function and metabolism of insect muscles in relation to their structure. Quart. Journ. Micr. Sci., 96: 151–159, 8 figs.

Eggers, F. 1923. Ergebnisse von Untersuchungen am Johnstonchen Organ der Insekten und ihre Bedeutung für die allgemeine Beurteilung der stiftführenden Sinnesorgane. Zool. Anz., 57: 224–240, 2 figs.

Evenius, Christa. 1926. Der Verschluss zwischen Vorder- und Mitteldarm bei der postembryonalen Entwicklung von Apis mellifica L. Zool. Anz., 68: 249–262, 10 figs.

Evenius, Christa. 1933. Über die Entwicklung der Rektaldrüsen von Vespa vulgaris. Zool. Jahrb., Anat., 56: 349–372, 18 figs.

Evenius, J. 1925. Die Entwicklung des Zwischendarmes der Honigbiene. Zool. Anz., 63: 49–64, 7 figs.

REFERENCES

Evenius, J. and Christa. 1925. Kryptenzellen und Epithelregeneration im Mitteldarm der Honigbiene (*Apis mellifica* L.). Zool. Anz., 62: 250–256, 2 figs.

Flanders, S. E. 1950. Control of sex in the honeybee. Scientific Monthly, 71: 237–240, 1 fig.

Fraenkel, G., and Herford, G. V. B. 1938. The respiration of insects through the skin. Journ. Exp. Biol., 15: 266–280.

Freudenstein, K. 1928. Das Herz und das Circulationssystem der Honigbiene. Zeitschr. wiss. Zool., 132: 404–475, 23 figs.

Frings, H. 1944. The loci of olfactory end-organs in the honey-bee, *Apis mellifera* Linn. Journ. Exp. Zool., 97: 123–134, 1 fig.

Frisch, K. von. 1921. Über den Sitz der Geruchsinnes bei Insekten. Zool. Jahrb., allg. Zool., 38: 449–516, 7 text figs., 2 pls.

Frisch, K. von. 1948. Aus dem Leben der Bienen, 196 pp., 112 figs. Vienna.

Frisch, K. von. 1950. Die Sonne als Kompass im Leben der Bienen. Experimentia, 6(6): 210–221, 20 figs.

Fyg, W. 1932. Untersuchungen über Kalkkörperchen im Bienendarm. Schweiz. Bienen-Zeitung, Jahrg. 1932, No. 4: 1–5, 5 figs.

Fyg, W. 1942. Das Bienenblut. Schweiz. Bienen-Zeitung, n.s. 65: 120–122, 2 figs.

Fyg, W. 1952. The process of natural mating in the honeybee. Bee World, 33(8): 129–139, 5 figs.

Gerould, J. H. 1933. Orders of insects with heart-beat reversal. Biol. Bull., 64: 424–431.

Grandi, Marta. 1950. Contributi allo studio degli "Efemeroidei" Italiani. XIV. Boll. Ist. Ent. Univ. di Bologna, 18: 58–92, 21 figs.

Hagan, H. R. 1951. Embryology of the Viviparous Insects, 272 pp., 160 figs. New York.

Hanan, B. B. 1955. Studies on the retrocerebral complex in the honey bee. Part I: Anatomy and histology. Ann. Ent. Soc. America, 48: 315–320, 4 figs.

Hassanein, M. H. 1953. Studies on the normal and pathological histology of the alimentary canal of the honey-bee, *Apis mellifica* L. Bull. Soc. Fouad Ier d'Ent., 37: 345–357, 10 figs.

Haydak, M. H. 1943. Larval food and development of castes in the honeybee. Journ. Econ. Ent., 36: 778–792.

Hering, M. 1939. Die peritrophischen Hüllen der Honigbiene mit besonderer Berücksichtigung der Zeit während der Entwicklung des imaginalen Darmes. Ein Beitrag zum Studium der peritrophischen Membran der Insekten. Zool. Jahrb., Anat., 66: 129–190, 29 figs.

REFERENCES

Hertig, M. 1923. The normal and pathological histology of the ventriculus of the honeybee, with especial reference to infection with *Nosema apis*. Journ. Parasitol., 9: 109–140, 3 pls.

Heselhaus, F. 1922. Die Hautdrüsen der Apiden und verwandten Formen. Zool. Jahrb., Anat., 43: 369–464, 11 pls.

Hesse, R. 1901. Untersuchungen über die Organe der Lichtempfindung bei niederen Thieren. VII. Von den Arthropoden-Augen. Zeitschr. wiss. Zool., 70: 347–473, 6 pls.

Hodgson, E. S., Lettvin, J. Y., and Roeder, K. D. 1955. Physiology of a primary chemoreceptor unit. Science, 122: 417–418, 2 figs.

Hollande, A. C. 1914. Les cérodécytes ou oenocytes des insectes. Arch. Anat. Micr., 16: 1–66.

Hsü, F. 1938. Étude cytologique et comparée sur les sensilla des insectes. La Cellule, 47, 60 pp., 5 pls.

Inglesent, H. 1940. Zymotic function of the pharyngeal, thoracic and postcerebral glands of *Apis mellifica*. Biochem. Journ., 34: 1415–1418.

Jacobs, W. 1924. Das Duftorgan von *Apis mellifica* und ähnliche Hautdrüsenorgane sozialer und solitarer Apiden. Zeitschr. Morph. Ökol. Tiere, 3: 1–80, 27 figs.

Janet, C. 1911. Sur l'existence d'un organe chordotonal et d'une vesicule pulsatile antennaire chez l'abeille er sur la morphologie de la tête de cette espece. C. R. Acad. Sci. Paris, 152: 110–112, 1 fig.

Johannsen, O. A., and Butt, F. H. 1941. Embryology of Insects and Myriapods, 462 pp., 370 figs. New York and London.

Johnston, C. 1855. Auditory apparatus of the *Culex* mosquito. Quart. Journ. Micr. Sci., 3: 97–102.

Jonescu, C. N. 1909. Vergleichende Untersuchungen über das Gehirn der Honigbiene. Jenaische Zeitschr. Naturwis., 45: 111–180, 5 pls.

Jucker-Piédallu, Andrée. 1934. Anatomie de l'abeille adulte, 198 pp., 83 figs. St. Aubin (Neuchâtel).

Kalmus, H., and Ribbands, C. R. 1952. The origin of the odours by which honeybees distinguish their companions. Proc. R. Soc., London, B, 140: 50–59.

Kenyon, F. C. 1896. The brain of the bee. Journ. Comp. Neurol., 6: 133–210, 9 pls.

Kerr, W. E. 1951. Sex-chromosome in honey-bee. Evolution, 5(1): 80–81.

Koehler, Adrienne. 1920. Über die Einschlüsse der Epithelialzellen der Bienendarms und die damit in Beziehung stehenden Probleme der Verdauung. Zeitschr. angew. Ent., 7: 68–91, 3 figs.

Koehler, Adrienne. 1921. Beobachtungen über Veranderungen am Fettkörper der Biene. Schweiz. Bienen-Zeitung, 44: 424–428.

REFERENCES

Koschevnikov, G. A. 1900. Über den Fettkorper und die Oenocyten der Honigbiene (*Apis mellifera*, L.). Zool. Anz., 23: 337–353.

Kramer, S., and Wigglesworth, V. B. 1950. The outer layers of the cuticle in the cockroach *Periplaneta americana* and the function of the oenocytes. Quart. Journ. Micr. Sci., 91: 63–72, 1 pl.

Kratky, E. 1931. Morphologie und Physiologie der Drüsen in Kopf und Thorax der Honigbiene. Zeitschr. wiss. Zool., 139: 120–200.

Kusmenko, S. 1940. Über die postembryonalie Entwicklung des Darmes der Honigbiene und die Herkunft der larvalen peritrophischen Hüllen. Zool. Jahrb., Anat., 66: 463–530, 12 text figs., 4 pls.

Kusmenko, S. 1941. Herkunft der Malpighischen Gefässe der Honig-biene. Zool. Jahrb., Anat., 67: 271–292, 2 pls.

Laidlaw, H. 1944. Artificial insemination of the queen bee (*Apis mellifera* L.): Morphological bases and results. Journ. Morph., 74: 429–465, 15 figs.

Leuenberger, F. 1927. Afterköniginnen. Schweiz. Bienen-Zeitung, Jahrg. 1927, 3 pp., 3 figs.

Lewke, J. 1950. Zur Kenntnis der Wachsdrüsen der Honigbiene. Anat. Anz., 97: 265–268.

L'Hélias, Colette. 1950. Étude des glandes endocrines post-cérébrales de la larve d'*Apis mellifica*. Bull. Soc. Zool. France, 75: 70–74, 3 figs.

L'Hélias, Colette. 1952a. Étude des glandes endocrines post-cérébrales et du cerveau de la larve des *Lophyrus pini* (L.) et *rufus* (André) (Hyménoptères). Bull. Soc. Zool. France, 77: 106 112, 1 figs.

L'Hélias, Colette. 1952b. Étude de la glande prothoracique chez la larve d'*Apis mellifica* (Hyménoptère). Bul. Soc. Zool. France, 77: 191–195, 2 figs.

Löschel, F. 1916. Die postemtryonale Entwicklung des Geschlechtsapparates der Bienenkönigen (*Apis mellifica* L.). Zeitschr. angew. Ent., 3: 21–44, 3 text figs., 2 pls.

Lotmar, Ruth. 1945. Die Metamorphose des Bienendarms (*Apis mellifica*). Beihefte Schweiz. Bienen-Zeitung, 1(10): 443–506, 29 figs., 2 tables.

Machatschke, J. W. 1936. Der cuticuläre Afbau des Rhabdoms im Arthropodenauge. Mém. Soc. Zool. Tchécoslovaque de Prague, 4: 90–109, 2 pls.

Mackensen, O. 1943. The occurrence of parthenogenetic females in some strains of honeybees. Journ. Econ. Ent., 36: 465–467.

Manley, R. O. B. 1948. Workers mating. American Bee Journ., 88: 305.

Manning, F. J. 1949a. Sex-determination in the honey bee. The Microscope, May–June, 1949, 5 pp., 3 figs.

REFERENCES

Manning, F. J. 1949b. Sex-determination in the honey bee. II. Oogenesis. The Microscope, July–Aug., 1949, 2 pp., 1 fig.

Manning, F. J. 1949c. Sex-determination in the honey bee. III. Maturation of the egg. The Microscope, Sept.–Oct., 1949, 4 pp., 3 figs.

McIndoo, N. E. 1914a. The scent-producing organ of the honeybee. Proc. Acad. Nat. Sci., Philadelphia, 66: 542–555, 1 text fig., 2 pls.

McIndoo, N. E. 1914b. The olfactory sense of the honey bee. Journ. Exp. Zool., 16: 265–346, 24 figs.

McIndoo, N. E. 1922. The auditory sense of the honey-bee. Journ. Comp. Neurol., 34: 173–199, 26 figs.

McIndoo, N. E. 1939. Segmental blood vessels of the American cockroach (*Periplaneta americana*). Journ. Morph., 65: 323–351, 6 text figs., 2 pls.

McIndoo, N. E. 1945. Innervation of insect hearts. Journ. Comp. Neur., 83: 141–155, 4 figs.

Meier, K. 1916. Die postembryonale Entwicklung des Geschlechtsapparates der Arbeitsbiene (*Apis mellifica* L.). Zeitschr. angew. Ent., 3: 45–74, 3 text figs., 2 pls.

Menzer, G., and Stockhammer, K. 1951. Zur Polarisationoptik der Fazettenaugen von Insekten. Naturwissenschaften, 38: 190–191, 3 figs.

Metalnikoff, S., and Toumanoff, C. 1930. Les cellules sanguines et la phagocytose chez les larves d'abeilles. C. R. Soc. Biol., Paris, 103: 965–967, 1 fig.

Metzer, C. 1910. Die Verbindung zwischen Vorder- und Mitteldarm bei der Biene. Zeitschr. wiss. Zool., 96: 539–571, 2 pls.

Meves, F. 1907. Die Spermatocytenteilungen bei der Honigbiene (*Apis mellifica* L.) nebst Bemerkungen über Chromatenreduction. Archiv mikrosk. Anat. u. Entwicklungsmechanique, 70: 414–491, 5 pls.

Michaëlis, G. 1900. Bau und Entwicklung des männlichen Begattungsapparates der Honigbiene. Zeitschr. wiss. Zool., 67: 439–460, 1 pl.

Morison, G. D. 1927. The muscles of the adult honey-bee (*Apis mellifera* L.). Part I. Quart. Journ. Micr. Sci., 71: 395–463, 12 figs.

Morison, G. D. 1928a. The muscles of the adult honey-bee (*Apis mellifera* L.). Part II. Quart. Journ. Micr. Sci., 71: 563–651, 41 figs.

Morison, G. D. 1928b. The muscles of the adult honey-bee (*Apis mellifera* L.). Part III. Quart. Journ. Micr. Sci., 72: 511–526, 1 fig.

Nabert, A. 1913. Die Corpora allata der Insekten. Zeitschr. wiss. Zool., 104: 181–358, 8 text figs., 5 pls.

Nelson, J. A. 1915. The Embryology of the Honey Bee, 282 pp., 95 text figs., 6 pls. Princeton.

Nelson, J. A. 1917. The relation of the Malpighian tubules of the hind intestine in the honeybee larva. Science, n.s. 46: 343–345.

REFERENCES

Nelson, J. A. 1918. The segmentation of the abdomen of the honey-bee (*Apis mellifica* L.). Ann. Ent. Soc. America, 11: 1–8, 2 text figs., 1 pl.

Nelson, J. A. 1924. Morphology of the honeybee larva. Journ. Agr. Res., 28(12): 1167–1213, 5 text figs., 8 pls.

Nelson, J. A., Sturtevant, A. P., and Lineburg, B. 1924. Growth and feeding of honeybee larvae. U.S. Dept. Agric., Bull. 1222, 37 pp., 11 figs.

Newton, H. C. F. 1931. On the so-called olfactory pores in the honey bee. Quart. Journ. Micr. Sci., 74: 647–668, 5 text figs., 2 pls.

Nixon, H. L., and Ribbands, C. R. 1952. Food transmission within the honeybee community. Proc. R. Soc., London, B, 140: 43–50.

Nowikoff, M. 1931a. Untersuchungen über die Komplexaugen von Lepidopteren nebst einigen Bemerkungen über die Rhabdome der Arthropoden im allgemeinen. Zeitschr. wiss. Zool., 138: 1–67, 4 pls.

Nowikoff, M. 1931b. Das Modell des Rhabdoms von Komplexaugen. Biol. Zentralb., 51: 325–329, 1 fig.

Nutting, W. L. 1951. A comparative anatomical study of the heart and accessory structures of the orthopteroid insects. Journ. Morph., 89: 501–598, 21 pls.

Oertel, E. 1930. Metamorphosis in the honey bee. Journ. Morph., 50: 295–340, 4 pls.

Örösi-Pál, Z. 1932. Das Verhalten der eierlegenden Arbeitsbiene. Zool. Anz., 98: 259–267, 2 figs.

Palm, N. B. 1949. The pharyngeal gland in *Bombus* Latr. and *Psithyrus* Lep. Opuscula Ent. (Lund), 14: 27–47, 8 figs.

Park, O. W. 1925. The storing and ripening of honey by honeybees. Journ. Econ. Ent., 18: 405–410, 2 pls.

Park, O. W. 1954. How bees make honey. American Bee Journ., 94: 296–298, 309.

Parker, R. L. 1926. The collection and utilization of pollen by the honeybee. Cornell Univ. Agric. Exp. Sta., Mem. 98, 55 pp., 16 figs.

Paulcke, W. 1900. Über die Differenzierung der Zellelemente im Ovarium der Bienenkönigen (*Apis mellifica* ♀). Zool. Jahrb., Anat., 14: 177–202, 4 pls.

Pavlovsky, E N., and Zarin, E. J. 1922. On the structure of the alimentary canal and its ferments in the bee (*Apis mellifera*, L.). Quart. Journ. Micr. Sci., 66: 509–556, 3 pls.

Pérez, C. 1903. Contribution à l'étude des metamorphoses des insectes. Bull. Sci. France et Belgique, 37: 195–427, 3 pls.

Pflugfelder, O. 1937. Bau, Entwicklung und Funktion der Corpora allata und cardiaca von *Dixippus morosus* Br. Zeitschr. wiss. Zool., 149: 477–512, 26 figs.

REFERENCES

Pflugfelder, O. 1948. Volumetrische Untersuchungen an den Corpora allata der Honigbiene, *Apis mellifica* L. Biol. Zentralb., 67: 223–241, 13 figs.

Phillips, E. F. 1905. Structure and development of the compound eye of the honeybee. Proc. Acad. Nat. Sci., Philadelphia, 57: 123–157, 3 pls.

Pringle, J. W. S. 1938a. Proprioception in insects. I. A new type of mechanical receptor from the palps of the cockroach. Journ. Exp. Biol., 15: 101–113, 8 figs.

Pringle, J. W. S. 1938b. Proprioception in insects. II. The action of the campaniform sensilla on the legs. Journ. Exp. Biol., 15: 114–131, 13 figs.

Redikorzew, W. 1900. Untersuchungen über den Bau der Ocellen der Insekten. Zeitschr. wiss. Zool., 68: 581–624, 7 text figs., 2 pls.

Rehm, E. 1939. Die Innervation der inneren Organe von *Apis mellifica*. Zeitschr. Morph. Ökol. Tiere, 36: 89–122, 26 figs.

Reimann, K. 1952. Neue Untersuchungen über die Wachsdrüse der Honigbiene. Zool. Jahrb., Anat., 72: pp. 251–272, 3 pls.

Rengel, C. 1903. Über den Zusammenhang von Mittledarm und Enddarm bei den Larven der aculeaten Hymenopteren. Zeischr. wiss. Zool., 75: 221–232, 2 pls.

Ribbands, R. 1953. The Behaviour and Social Life of Honeybees, 352 pp., 66 text figs., 9 pls. London.

Ribbands, R. 1955. The origin of bee scents. American Bee Journ., 95: 270–271.

Richard, G. 1952. L'innervation sensorielle pendant les mues chez les insectes. Bull. Soc. Zool. France, 77: 99–106, 3 text figs., 1 pl.

Richards, A. G. 1951. The Integument of Insects, 411 pp., 65 figs. Minneapolis.

Richards, A. G. 1952. Studies on arthropod cuticle. VIII. The antennal cuticle of honeybees, with particular reference to the sense plates. Biol. Bull., 103: 201–225, 46 figs.

Richards, A. G., and Korda, Frances H. 1950. Studies on arthropod cuticle. IV. An electron microscope survey of the intima of arthropod tracheae. Ann. Ent. Soc. America, 43: 49–71, 3 pls.

Rietschel, P. 1937. Bau und Funktion des Wehrstachels der staatenbildenden Bienen und Wespen. Zeitschr. Morph. Ökol. Tiere, 33: 313–357, 20 figs.

Roberts, W. C. 1944. Multiple mating of queen bees proved by progeny and flight tests. Glean. Bee Cult., 72: 255–259.

Roonwal, M. L. 1936. Studies on the embryology of the African migratory locust, *Locusta migratoria migratorioides* R. and F. I. The early development. Philosoph. Trans. R. Soc. London, B, No. 538, 226: 391–421, 16 text figs., 3 pls.

Roonwal, M. L. 1937. Studies on the embryology of the African migratory locust, *Locusta migratoria migratorioides* R. and F. II. Organogeny. Philosoph. Trans. R. Soc. London, B, No. 543, 227: 175–244, 7 pls.

Root, A. I. and E. R. 1917. The mating of queen and drone. The ABC and XYZ of Bee Culture, pp. 229–231.

Rösch, G. A. 1927a. Über die Bautätigkeit im Bienenvolk und das Alter der Baubienen. Zeitschr. vergl. Physiol., 6: 264–298, 10 figs.

Rösch, G. A. 1927b. Beobachtungen an Kittharz sammelnden Bienen (*Apis mellifica* L.). Biol. Zentralb., 47: 113–121, 2 figs.

Rösch, G. A. 1930. Untersuchungen über die Arbeitsteilung im Bienenstaat. 2. Teil: Die Tätigkeiten der Arbeitsbienen unter experimentell veränderten Bedingunen. Zeitschr. vergl. Physiol., 12: 1–71, 25 figs.

Ross, H. H. 1936. The ancestry and wing venation of the Hymenoptera. Ann. Ent. Soc. America, 29: 99–111, 2 pls.

Ruttner, F. 1956. Zur Frage der Spermaübertragung bei der Bienenkönigin. Insectes Sociaux (in press).

Ruttner, S. and F. 1953, 1954. Über die Paarung der Bienenkönigin. Osterreichische Imker, 3: 206–211, 4: 12–14, 27–30.

Schaller, F. 1950. Étude morphologique du complexe endocrine rétrocérébral de la larve d'abeille (*Apis mellifica* L.). C. R. Soc. Biol. Paris, 144: 1097–1100, 5 figs.

Schaller, F. 1951. Réalisation des charactères des caste au cours de dével oppements perturbés chez l'abeille (*Apis mellifica* L.). C. R. Soc. Biol. Paris, 145: 1351–1354, 2 figs

Schaller, F. 1952. Effets d'une ligature postcéphalique sur le développement de larves agées d'*Apis mellifica* L. Bull. Soc. Zool. France, 77: 195–204, 9 figs.

Scharrer, E., and Scharrer, B. 1954. Hormones produced by neurosecretory cells. Recent Progress in Hormone Research, Proc. Laurentian Hormone Conference, 10: 183–240, 14 figs.

Schenk, O. 1903. Die antennalen Hautsinnesorgane einiges Lepidopteren und Hymenopteren. Zool. Jahrb., Anat., 17: 573–618, 4 text figs., 2 pls.

Schiemenz, P. 1883. Über das Herkommen des Futtersaftes und die Speicheldrüsen der Biene nebst einem Anhange über das Riechorgan. Zeitschr. wiss. Zool., 38: 71–135, 3 pls.

Schmieder, R. G. 1928. Observations on the fat-body in Hymenoptera. Journ. Morph., 45: 121–184.

Schnelle, H. 1923. Über den feineren Bau des Fettkörpers der Honigbiene. Zool. Anz., 57: 172–179, 1 fig.

Schön, A. 1911. Bau und Entwicklung der tibialen Chordotonalorgane bei der Honigbiene und bei Ameisen. Zool. Jahrb., Anat., 31: 439–472.

Schreiner, T. 1952. Über den Nahrungstransport im Darm der Honibgiene. Zeitschr. vergl. Physiol., 34: 278–298, 2 figs.

Shafer, G. D. 1917. A study of the factors which govern mating in the honeybee. Michigan Agr. Col. Exp. Sta., Div. of Ent., Technical Bull. 34, 19 pp., 3 pls.

Short, J. R. T. 1952. The morphology of the head of larval Hymenoptera with special reference to the head of Ichneumonoidea, including a classification of the final instar larvae of Braconidae. Trans. R. Ent. Soc. London, 103: 27–84, 34 figs.

Sihler, H. 1924. Die Sinnesorgane an der Cerci der Insekten. Zool. Jahrb., Anat., 45: 519–580, 4 pls.

Sikes, Enid K., and Wigglesworth, V. B. 1931. The hatching of insects from the egg, and the appearance of air in the tracheal system. Quart. Journ. Micr. Sci., 74: 165–192, 8 figs.

Sladen, F. W. L. 1911. How pollen is collected by the social bees, and the part played in the process by the auricle. British Bee Journ., 39: 491–514, 4 figs.

Snodgrass, R. E. 1910. The Anatomy of the Honey Bee, 162 pp., 57 figs. U.S. Dept. Agric., Bur. Ent., Tech. Series, No. 18.

Snodgrass, R. E. 1925. Anatomy and Physiology of the Honeybee, 327 pp., 108 figs. New York and London.

Snodgrass, R. E. 1935. Principles of Insect Morphology, 667 pp., 319 figs. New York and London.

Snodgrass, R. E. 1941. The male genitalia of Hymenoptera. Smithsonian Misc. Coll., 99(14), 86 pp., 33 pls.

Snodgrass, R. E. 1942. The skeleto-muscular mechanisms of the honey bee. Smithsonian Misc. Coll., 103(3), 120 pp., 32 figs.

Sotavalta, O. 1947. The flight-tone (wing-stroke frequency) of insects. Acta Entomologica Fennica, 4: 1–117.

Soudek, S. 1927. The pharyngeal glands of the honeybee (Apis mellifica L.). Bull. de l'École Sup. d'Agron. Brno. Sign. C10: 1–61, 6 text figs., 3 pls.

Stellwaag, F. 1910. Bau und Mechanik des Flugapparates der Biene. Zeitschr. wiss. Zool., 95: 518–550, 6 text figs., 2 pls.

Straus, J. 1911. Die chemische Zusammensetzung der Arbeitsbienen und Drohnen während ihrer verschiedenen Entwicklungstadien. Zeitschr. Biologie, 56: 347–397.

Terre, L. 1900. Contribution à l'étude de l'histolyse du corps adipeux chez l'abeille. Bull. Soc. Ent. France, 1900: 62–66.

REFERENCES

Thomsen, M. 1954. Neurosecretion in some Hymenoptera. Kongel. Danske Vidensk. Selskab. Biol. Skr., 7(5), 24 pp., 8 pls.

Thorpe, W. H. 1949. Orientation and methods of communication of the honey bee and its sensitivity to the polarization of the light. Nature, 164: 11–19.

Tiegs, O. W. 1955. The flight muscles of insects—their anatomy and histology; with some observations on the structure of striated muscle in general. Philosoph. Trans. R. Soc. London, B, No. 656, 238: 221–347, 17 text figs., 16 pls.

Tirelli, M. 1929. Sbocco dei canali malpighiani nell'intestino medio. Boll. Mus. Zool. Anat. Comp. della R. Univ. Genova, 9: 1–18, 3 figs.

Trappmann, W. 1923. Die Malpigische Gefässe, Anatomie und Physiologie des Zwischendarms, Die Bildung der peritrophischen Membran, Die Rictaldrüsen von *Apis mellifica* L. Arch. Bienenkunde., 5: 177–220, 20 figs.

Trojan, E. 1930. Die Dufoursche Drüse bei *Apis mellifica*. Zeitschr. Morph. Ökol. Tiere, 19: 678–685, 3 figs.

Vejdovsky, F. 1925. Quelques remarques sur la structure et le développement des cellules adipeuses et des oenocytes pendant la nymphose de l'abeille. La Cellule, 35: 63–103.

Vowles, D. M. 1955. The structure and connections of the corpora pedunculata in bees and ants. Quart. Journ. Micr. Sci., 96: 239–255, 9 figs.

Watanabe, Mary I., and Williams, C. M. 1951. Mitochondria in the flight muscles of insects. I. Chemical composition and enzymatic content. Journ. Gen. Physiol., 34: 675–689, 5 figs.

Watanabe, Mary I., and Williams, C. M. 1953. Mitochondria in the flight muscles of insects. II. Effects of the medium on the size, form, and organization of isolated sarcosomes. Journ. Gen. Physiol., 37: 71–90, 5 text figs., 1 pl.

Waterhouse, D. F. 1953. The occurrence and significance of the peritrophic membrane, with special reference to adult Lepitoptera and Diptera. Australian Journ. Zool., 1: 299–318, 2 pls.

Weaver, N. 1955. Rearing of honeybee larvae on royal jelly in the laboratory. Science, 121: 509–510.

Weber, H. 1933. Lehrbuch der Entomologie, 726 pp., 556 figs. Jena.

Weber, H. 1954. Grundriss der Insektenkunde, 3d ed., 428 pp., 220 figs. Stuttgart.

Weil, E. 1935. Vergleichend-Morphologische Untersuchungen am Darmkanal einiger Apiden und Vespiden. Zeitschr. Morph. Ökol. Tiere, 30: 438–478, 24 figs.

Weyer, F. 1935. Über drüsenartige Nervenzellen im Gehirn der Honigbiene *Apis mellifica* L. Zool. Anz., 112: 137–141, 3 figs.

Wheeler, W. M. 1923. Social Life among the Insects, 375 pp., 116 figs. New York.

Whitcomb, W. Jr., and Wilson, H. F. 1929. Mechanics of digestion of pollen by the adult honey bee and the relation of undigested parts to dysentery of bees. Agric. Exp. Sta., Univ. of Wisconsin, Research Bull. 92, 27 pp., 11 figs.

White, G. F. 1918. A note on the muscular coat of the ventriculus of the honey bee (*Apis mellifica*). Proc. Ent. Soc. Washington, 20: 152–154, 1 fig.

Wigglesworth, V. B. 1930. A theory of tracheal respiration in insects. Proc. R. Soc., London, B, 106: 229–250, 10 figs.

Wigglesworth, V. B. 1932. On the function of the so-called 'rectal glands' of insects. Quart. Journ. Micr. Sci., 75: 131–150.

Wigglesworth, V. B. 1942. The storage of protein, fat, glycogen and uric acid in the fat body and other tissues of mosquito larvae. Journ. Exp. Biol., 19: 56–77, 11 figs.

Wigglesworth, V. B. 1953. The origin of sensory neurones in an insect, *Rhodnius prolixus* (Hemiptera). Quart. Journ. Micr. Sci., 94: 93–112, 16 figs.

Willson, R. B. 1955. Royal jelly: a review. American Bee Journ., 95: 15–21.

Wolff, O. J. B. 1875. Das Riechorgan der Biene. Nova Acta, K. Leop.-Carol. Deutschen Akad. Naturf., 38, 251 pp., 8 pls.

Wolsky, A. 1931. Weitere Beiträge zum Ocellenproblem. Die optischen Verhaltnisse der Ocellen der Honigbiene (*Apis mellifica* L.). Zeitscher. vergl. Physiol., 14: 385–414, 2 figs.

Woyke, J. 1955. Multiple mating of the honeybee queen (*Apis mellifica* L.) in one nuptial flight. Bull. Acad. Polonaise Sci., Cl. II, 3, No. 5: 175–180, 8 figs.

Zander, E. 1900. Beiträge zur Morphologie der männlichen Geschlechtsorgane der Hymenopteren. Zeitschr. wiss. Zool., 47: 461–489, 9 text figs., 1 pl.

Zander, E. 1910. Die Gliederung des thoracalen Hautskelletes der Bienen und Wespen. Zeitschr. wiss. Zool., 95: 507–517, 8 text figs., 1 pl.

Zander, E. 1911. Der Bau der Biene, 182 pp., 149 text figs., 20 pls. Stuttgart.

Zander, E. 1916. Die Ausbildung des Geschlechtes bei der Honigbiene (*Apis mellifica* L.). Zeitschr. angew. Ent., 3: 1–20, 2 pls.

INDEX